D1454021

Statistics of the Boolean Model for Practitioners and Mathematicians

Statistics of the Boolean Model for Practitioners and Mathematicians

Ilya Molchanov
University of Glasgow

JOHN WILEY& SONS
Chichester • New York • Weinheim • Brisbane • Singapore • Toronto

Other Wiley Editorial Offices

John Wiley & Sons, Inc., 605 Third Avenue,
New York, NY 10158-0012, USA

VCH Verlagsgesellschaft mbH, Pappelallee 3,
D-69469 Weinheim, Germany

Jacaranda Wiley Ltd, 33 Park Road, Milton,
Queensland 4064, Australia

John Wiley & Sons (Asia) Pte Ltd, 2 Clementi Loop #02-01,
Jin Xing Distripark, Singapore 0512

John Wiley & Sons (Canada) Ltd, 22 Worcester Road,
Rexdale, Ontario M9W 1L1, Canada

ISBN 0 471 97102 2

British Library Cataloguing in Publication Data

A catalogue record for this book is available from the British Library

Produced from camera-ready copy supplied by the author
Printed and bound in Great Britain by Biddles Ltd, Guildford and King's Lynn
This book is printed on acid-free paper responsibly manufactured from sustainable
forestation, for which at least two trees are planted for each one used for paper production.

Contents

1

Introduction

The ability to see appeared in the history of mankind much earlier than the ability to count numbers, and the very first 'statistical' experiences dealt with average shapes of things that were seen. For example, man encountered the evaluation of an 'average impression' of tigers or sheep before starting to count the numbers seen.

In contrast, the development of mathematically strict statistical methods followed the reverse order. Statistics began with the analysis of series of numbers. Later on, statistical methods were extended to the study of random processes (time series) in order to describe an evolution of random variables as time proceeds. However, the statistical analysis of pictures, graphic patterns, and images remained untouched for a long time. Nowadays it is treated by two branches of statistics: statistical image analysis on the one hand and stochastic geometry and spatial statistics on the other.

Statistical image analysis and pattern recognition in their classical form (in relation to the analysis of pictures) deal with classification, structural analysis, storage, etc. of a single picture (perhaps, spoiled or modified by random noise), see [Bad86, Pos91, Pra77] and references therein. The relevant technique is based on classification theory [Fuk72], Markov random fields [Bes86], Gibbs random fields [GG84] and Markov chain Monte Carlo [Gey93, GRS96], Fourier transforms [DH73], mathematical morphology [Hei94a, Hei95], etc.

Spatial statistics works with series of essentially random pictures and aims to find out probability laws governing these random images. It is also important to determine average numerical values, which characterise the image in a certain sense. The ultimate goal is to estimate parameters of a relevant mathematical model and test its applicability and appropriateness.

The situation can be explained by the following simple example. Suppose that we have a photograph of an object in a room. The purpose of pattern recognition theory in this framework could be to classify and recognise this object, for example, to determine if it is a chair or a table. The answer can be given in a probability form: this is a chair with probability p and a table with probability $1-p$. Sometimes related problems are also treated: to remove noise, to enhance the picture, to find an optimal way of storing this image, etc.

There are two main branches of statistics of random sets. The first deals with statistical analysis of independent identically distributed random compact sets. In the 'chair–table' language it can be explained as finding the 'expected' chair, or fitting a model of 'random chairs' if a sample of chairs is given. The main problem here is the lack of information about original orientations and locations of the sets considered. Then, it is more appropriate to speak about statistics of figures rather than about statistics of sets. Although for chairs a kind of 'more preferable orientation' is possible to be determined, this is useless if a sample of sand or diamond grains is considered, see [SS94, SM97, Zie89, Zie94].

We will now deal with the other part of statistics of random sets, which treats stationary random sets. Suppose that we have a photograph of a room full of similar objects of different sizes and in different positions (a 'random warehouse' without regular shelves). Then spatial statistics can help to decide if this is a warehouse of tables or of chairs. The answer could be also given in the form of a probability law, say a randomly chosen (typical) object is a chair with probability p or a table with probability $1-p$. Other questions are related to the evaluation of the 'mean chair' or the distribution of the size or shape of the 'typical chair'. In this case a kind of spatial homogeneity (and sometimes isotropy) is assumed. In other words, a picture (set) is considered to be a part of a larger stationary set of 'mathematically' infinite size. This problem is similar to time series analysis in classical statistics. Again, the 'chair' language gives such an example if the 'random' warehouse of chairs is sufficiently large.

Certainly, there are much more interesting and useful examples than 'random warehouses' of chairs and tables. They appear when a picture is produced from similar elementary components placed randomly in space. For example, it can depict red blood particles in the viewing field of a microscope, an area occupied by forest or grass, a system of water droplets, fibres of paper or pores in cheese. Such a picture can be either imported from a microscope or copied from a geographical map.

If the picture to be analysed is partially destroyed or noisy, then it can be smoothed by using either morphological operations [Ser82, Hei94a, Hei95] (some graphic editors on computers perform this quite satisfactorily) or filtration theory for Markov random fields [Bes86, God91], which ranges from rather elementary methods to sophisticated techniques. Therefore, we will assume that our pictures are not distorted by either statistical or deterministic errors. We also consider mostly planar pictures. However, for some methods it is not difficult to change the dimension. We also put aside the stereological framework [Sto90, SKM95, Wei80], where lower-dimensional sections of a picture cause loss of information.

The main technical device that is used to perform the algorithms and methods of geometrical statistics is the so-called 'image analyser'. It can be viewed as a computer (with relevant software) attached to a source of images (video-camera, scanner, microscope, etc.). An image analyser must be able to import and store images and make calculations with them, for instance, to compute numbers of pixels (corresponding to the area covered), the boundary length, the minimum distance from a point to a set, etc. The successful performance of these (inherently continuous) operations on an inherently digitised computer screen is not in general an easy problem, see [Ser82, Chapter VII] and [Hei94a].

We consider only binary images. First, the corresponding mathematical techniques are simpler and better developed. On the other hand, a grey-tone image can be transformed to a family of binary pictures by means of a family of thresholds. Other (also very important) problems like practical implementation on image analysers, filtering of images, classification and image storage will not be considered at all.

This book discusses statistical models of stationary random sets, estimation of parameters and fitting appropriate stochastic models. The presentation begins with the definition of the Boolean model of random closed sets. This model emerges from the Poisson point process if the points are replaced by sets of 'full' dimension. The overlappings that arise constitute the main problem in statistical estimation theory and statistical inference. In fact, all model parameters can be classified as either aggregate or individual parameters. The first are determined directly by the visible set, while the latter are not directly observable and can be related to aggregate parameters through mathematical equations. Note that the estimation of aggregate parameters is similar to the estimation of time-average characteristics for time series. Then equations relating individual and aggregate parameters and the corresponding estimators are discussed and applied to statistical estimation of parameters. The corresponding problems in time series statistics are fitting an autoregression scheme or estimating the trend. After this, the parametric approach and more sophisticated sampling schemes are considered. Finally, some testing problems are briefly discussed.

The material given in this book is divided into two levels. The first consists of basic ideas, definitions, explanations, simple properties, descriptions of algorithms, and recipes of how to implement them. It deals only with the planar case and does not go deep into the mathematical background, although corresponding references are usually mentioned. For this level intuitive understanding will be mostly enough. The other part of the presentation (given as notes) assumes general dimension of the space, and contains some mathematical background, and occasionally proofs. It is *always* possible to skip the second part, if only implementations are of interest.

This book has emerged from a manuscript written for applied scientists and explaining how to do statistics of the Boolean model. This material will be used in the first level of presentation. However, the existence of the interesting mathematical background and the necessity of further developments were the reasons to add the second (mathematical) part aimed at statisticians and probabilitists. It is necessary to stress that statistics of the Boolean model is by no means complete. New ideas must appear to compare estimators and to extend for random sets (and Boolean models) most of the well-known methods of mathematical statistics.

I would like to thank Dietrich Stoyan for attention to this work, friendly critique and encouragement. Many colleagues supplied me with reprints and unpublished papers. I am especially grateful to Wolfgang Weil for explanations of integral-geometric methods.

I began this work at Freiberg University of Mining and Technology (TU Bergakademie Freiberg). I am indebted to my Freiberger colleagues for their help and kindness which made my stay there really pleasant and to the Alexander von Humboldt Foundation for the financial support.

I am grateful to the Department of Theoretical Statistics of Aarhus University for giving me the chance to give a short course on this subject, which provided a chance to polish the manuscript, and also to John Goutsias who kindly looked through some of these notes and gave many useful remarks.

My former visits to the Centre of Mathematics and Computer Science (CWI, Amsterdam) were always extremely useful and inspiring. Having spent almost two years there and having enjoyed this environment every day, I would like to express my gratitude to Adrian Baddeley for the invitations and many interesting discussions. This work has been partially supported by the project 'Computationally Intensive Statistical Methods' sponsored by the Netherlands Organisation for Scientific Research (NWO).

I am indebted to Jean-Louis Quenec'h who has supplied me with beautiful examples of Boolean models. Robin Pohlink from Freiberg University of Mining and Technology and Adri Steenbeek from CWI helped me a lot with the relevant software necessary to handle the pictures. Finally, I would like to thank my Glasgow colleagues for help and support that were crucial to the completion of this project.

Ilya Molchanov   Glasgow, September 1996

2

The Boolean Model

2.1 Basic Notation

Most mathematical notation is introduced as it is needed. Some generally accepted conventions follow below.

We will work with random sets in the Euclidean space $\mathbf{R}^d$. Points of $\mathbf{R}^d$ are denoted by small Latin letters, usually x, y, z, u, v, w. We consider mostly two-dimensional space $\mathbf{R}^2$, i.e. the dimension d is supposed to be 2. In general, the dimension d is explicitly mentioned unless the corresponding results are valid for any d.

For two points $x, y \in \mathbf{R}^d$ the Euclidean distance between them is denoted by

$$\rho(x, y) = \left(\sum_{i=1}^{d} (x_i - y_i)^2 \right)^{1/2} ,$$

where $x = (x_1, \ldots, x_d)$ and $y = (y_1, \ldots, y_d)$. In particular, $\|x\| = \rho(x, o)$ is the norm of x, where o is the origin. The ball of radius r centred at x is denoted by $B_r(x)$. In other words, $B_r(x) = \{y : \rho(x, y) \le r\}$ is the set of all y such that the distance between x and y is less than or equal to r. A unit ball centred at the origin is denoted by B. Furthermore, $\rho(x, K) = \inf\{\rho(x, y) : y \in K\}$ is the (minimum) distance between x and $K \subset \mathbf{R}^d$. The Minkowski addition of two sets in $\mathbf{R}^d$ is defined by $F_1 \oplus F_2 = \{x_1 + x_2 : x_1 \in F_1, x_2 \in F_2\}$.

For any planar set F, its area and perimeter (if they exist) are denoted by $\mathsf{A}(F)$ and $\mathsf{U}(F)$ respectively. In $\mathbf{R}^3$ we denote the volume of F by $\mathsf{V}(F)$. For general dimensions $\mathsf{mes}(F)$ (or $\mu_d(F)$) denotes the d-dimensional Lebesgue measure of F, i.e. $\mathsf{mes}(F) = \mathsf{A}(F)$ for $d = 2$ and $\mathsf{mes}(F) = \mathsf{V}(F)$ for $d = 3$. If the set F is finite, then $\mathsf{N}(F)$ denotes the number of points in F (the cardinality of F).

A set F in $\mathbf{R}^d$ is said to be closed if, for each convergent sequence $\{x_n\} \subset F$, its limit also belongs to F. A set G is open if its complement is closed. Equivalently, for each point x of G there exists $r > 0$ such that $B_r(x) \subset G$. Furthermore, $F \subset \mathbf{R}^d$ is bounded if

$$\|F\| = \sup\{\|x\| : x \in F\} < \infty,$$

or, equivalently, $F \subset B_R(o)$ for some $R > 0$. A set K is said to be compact if each sequence of its points contains a convergent subsequence. In $\mathbf{R}^d$ this condition is satisfied if and only if K is closed and bounded. The family of all closed (resp. compact) sets in $\mathbf{R}^d$ is denoted by $\mathcal{F}$ (resp. $\mathcal{K}$).

Some letter conventions are worth mentioning. Usually the letters t, s, r usually denote real numbers, i, j, m, n, l, k denote integers, x, y, z, u, v, w denote points in $\mathbf{R}^d$. The capitals F, K, G (also with sub- or super-scripts) denote some subsets of $\mathbf{R}^d$ (usually closed, compact and open respectively). Furthermore, X, Y, Z, Ξ are used to denote random closed sets.

2.2   Point Patterns and Point Processes

Point patterns constitute the visually simplest class of graphic images. Although every picture is composed of points, the notion 'point pattern' means that the corresponding set of points is locally finite, i.e. each bounded set contains at most a finite number of points from this pattern for the real (not digitised!) image. Roughly speaking, geometrical statistics began with statistical analysis of point patterns. In the course of time it has been applied to many practical problems, beginning with the analysis of red blood particles in [Abb78]. Other references and historical discussions can be found in [CI80, DVJ88] and [SKM95, Chapter 2].

A locally finite point configuration on the plane (or in a general space) can be determined by its points $\Psi = \{x_1, x_2, \ldots\}$. Another possibility is to take as known the number $\mathsf{N}(K)$ of points in each (measurable) bounded set K. A random locally finite point configuration is said to be a point process. The word 'process' in the term 'point process' does not mean a real process, i.e. a time-dependent evolution. However, some analogies are possible. For example, time averages in the theory of random processes can usually be replaced by spatial averages in the theory of point processes.

The simplest random point process is the stationary *Poisson point process* $\Pi_\lambda = \{x_1, x_2, \ldots\}$ in $\mathbf{R}^d$, whose distribution is determined by the following two conditions.

1. For each bounded (measurable) set K the number of points $\mathsf{N}(K)$ follows the Poisson distribution with parameter $\lambda \mathsf{mes}(K)$ (or $\lambda \mathsf{A}(K)$ in the planar case).

2. For disjoint sets $K_1, \ldots, K_n$ the numbers $\mathsf{N}(K_1), \ldots, \mathsf{N}(K_n)$ are independent.

The parameter λ is said to be the *intensity* of the point process and can be interpreted as the mean number of points of the process in a unit area. We consider mostly the planar stationary case. In general, $\lambda \mathrm{mes}(K)$ should be replaced by $\Lambda(K)$ for a locally finite (intensity) measure Λ. Then $\mathsf{N}(K)$ has the Poisson distribution with parameter $\Lambda(K)$.

The simulation of a Poisson point process within a bounded set K can be performed in the following way. First, it is necessary to simulate the Poisson random variable N with the parameter $\lambda \mathsf{A}(K)$. Then N independent points uniformly distributed within K give a realisation of the Poisson point process.

Already the Poisson point process can be applied to many practical problems; see, for example, [Dig83, Oga88, Pie77, SKM95]. However, Π_λ can be used to construct more sophisticated models of point processes, as in [SKM95, Chapter 5] and [BvL95].

It is possible to discern two main statistical problems related to the Poisson point process:

- estimating the intensity;
- testing the Poisson property.

Both have been studied extensively, see [Kar86, LL85, SKM95, UF85].

The Poisson point process considered here is stationary and ergodic. The latter means that the mean characteristics of the process can be obtained from spatial averages of the functionals of this process. However, it is not possible technically to observe a whole infinite realisation of this point process. An observer possesses only bounded observations made through a certain window W. We always assume the following framework for estimation problems:

> A stationary random pattern (point process, random closed set) is observed in a convex bounded (compact) window W. An estimator for some parameter constructed by observations in W is denoted by the same letter as the parameter with a 'hat' and index W. Consistency and limit theorems for estimators are understood as the window W expands to the whole plane (or space), i.e. $W \uparrow \mathbf{R}^d$.

In practice all estimates are computed for a fixed window W (for example, a display's screen). These expanding windows correspond to the classical statistical scheme of independent observations, where the number of observations tends to infinity, although in practice only a finite number of them are available. It is convenient to consider the given window W to be an element of the expanding family of windows W_s, $s > 0$. Also it is useful to have in mind the most common case $W_s = sW_1$ for a fixed convex set W_1 containing the origin as an interior point. We will always assume this.

In this framework the intensity of the Poisson point process Π_λ is estimated by

$$\hat{\lambda}_W = \frac{N(\Pi_\lambda \cap W)}{A(W)}, \tag{2.1}$$

where $N(\Pi_\lambda \cap W)$ is the number of points of Π_λ lying inside W. This is a strong consistent and efficient estimator of the intensity, that is, $\hat{\lambda}_W$ converges to λ with probability one as $W \uparrow \mathbf{R}^d$ and the variance of estimator (2.1) is the smallest for all unbiased estimators.

Further information and references on statistics of the Poisson point process can be found in [DVJ88, Kar86, SKM95].

Notes to Section 2.2

Point processes. A general point process Ψ can be viewed as a random element in the space of locally finite point configurations endowed with the corresponding σ-algebra. Another possible way is to define it as a random counting measure $N(\cdot)$, see [CI80, DVJ88, SKM95] for the exact definitions of measurability. This counting measure gives the number of points $N(\Psi \cap K)$ in a bounded Borel set K. Thus, the Poisson point process appears in the special case when the random measure $N(\cdot)$ satisfies the conditions given above.

In general, the expectation of $N(\cdot)$ is said to be the intensity measure of the corresponding point process. Stationarity means that the distribution of $N(\cdot)$ does not change after any non-random translation. This implies, in particular, that the intensity measure is proportional to the Lebesgue measure $\text{mes} = \mu_d$, i.e. $\mathbf{E}N(K) = \lambda\text{mes}(K)$. The proportionality coefficient λ is said to be the *intensity* of the point process.

A useful result from the theory of point processes is *Campbell's theorem* which states that

$$\mathbf{E}\left[\sum_x f(x)\right] = \mathbf{E}\left[\int f(x)N(dx)\right] = \int f(x)\mathbf{E}N(dx), \tag{2.2}$$

i.e. for each integrable measurable function f, the mean of the integral with respect to the counting measure is equal to the integral with respect to the intensity measure.

Some results from the theory of point processes will be used occasionally later on. To follow the proofs the reader is advised to look at the relevant pages in [DVJ88, KS91] and [SKM95]. A thorough presentation of statistics of point processes and many references can be found in [Kar86].

Poisson point processes on general spaces. Let E be a general locally compact separable complete space, and let θ be a σ-finite measure on E. In this case it is also possible to define the Poisson point process on E with intensity measure θ. Then the number of points in each bounded Borel set F has the Poisson distribution with parameter $\theta(F)$ and the numbers of points in disjoint sets are independent. In stochastic geometry it is usual to consider E to be a space of geometrical objects, for example, lines, planes, compact sets, cylinders, etc.

Expanding windows. The notation $W \uparrow \mathbf{R}^d$ means that the window of observations W expands to the whole space. It is convenient to view W as an element of a family W_s, $s > 0$, of windows. Usually the following assumptions are required [DVJ88, p. 332]

1. Each W_s is convex.
2. $W_s \subset W_t$ for $s < t$.
3. $\sup\{r : W_s$ contains a ball of radius $r\}$ tends to infinity as $s \to \infty$.

The simplest choice, $W_s = sW_1$, $s > 0$, satisfies these conditions for a convex compact set W_1 containing the origin. Further, the notation $W \uparrow \mathbf{R}^d$ always means that W is a member of a family of expanding windows growing in this regular way.

Interaction processes. A family of point processes can be defined via densities with respect to the Poisson point process. Typically, such a density has exponential form, e.g.,

$$f(\mathbf{x}) = \alpha\beta^{n(\mathbf{x})}\gamma^{\zeta(\mathbf{x})}, \quad \mathbf{x} \subset W,$$

where α, β and γ are positive constants, $\mathbf{x} = (x_1, \ldots, x_m)$ is a configuration of points inside a window W, $n(\mathbf{x}) = m$ is the number of points in the configuration, and $\zeta(\mathbf{x})$ is a certain functional depending on the geometry of $\mathbf{x}$. For example, the area-interaction point process [BvL95] is defined by taking

$$\zeta(\mathbf{x}) = \mathsf{mes}(Q(x_1) \cup \cdots \cup Q(x_m)),$$

where $Q(x_i)$ is a compact set in $\mathbf{R}^d$ determined by x_i, $1 \leq i \leq m$. Statistics of such point processes is based on using the so-called pseudo-likelihood approach [Bes78, Jen93, JM91] and methods based on simulations.

2.3   Definition of the Boolean Model

Let us look at the point pattern formed by a Poisson point process and suppose that the points are no longer points in the strict mathematical sense. If they are very small disks or squares, then their sizes can be neglected. However, if points are becoming really 'thick', then the picture takes on a different form. There appear overlappings and the area fraction (the part of the area covered) gets larger, so that for larger radii the whole plane can be covered. We assume that each point is 'thickened' independently of other points and of its own location.

This idea was used in the late 1960s by G. Matheron, J. Serra and their colleagues to define the Boolean model. For this, the points of the point process are replaced by independent random compact sets. The union of these random sets (grains), driven by the point process (germ points), serves as a model of the random set.

A *random compact set* can be viewed as a random element whose values are compact (i.e. closed and bounded) sets. One can always imagine a ball with a random radius or ellipses with random axes and orientations, see [Mat75, p. 27], [SKM95, p. 194] for rigorous definitions and [SS94, pp. 107–108] for a heuristic discussion. The random ball and the ellipse with random axes provide simple examples of random sets. In these particular cases the distribution of a random set can be reduced to a distribution of a random vector (for example, the coordinates of the random ball's centre and radius). In general it is not so.

Let us enlist several simple examples of random compact sets.

1. Random singleton $\{\xi\}$ or random finite-point set $\{\xi_1, \ldots, \xi_n\}$.
2. Random ball $B_\xi(o) = \xi B_1(o)$ of random radius ξ centred at the origin, or, more generally, the ball $B_\xi(\eta)$ with random centre η and random radius ξ.
3. ξF for a deterministic set F (the dilation of F by ξ).
4. Random rectangle with fixed or isotropic orientation.
5. Finite set $\{\xi_1, \ldots, \xi_n\}$ of random points sampled from a given distribution (e.g., uniformly distributed in a bounded set) or its convex hull $\mathrm{conv}\{\xi_1, \ldots, \xi_n\}$.
6. Random rotation of a deterministic set.
7. Poisson polygon generated by a field of random lines, see [Mat75, p. 168], [SKM95, pp. 324–326].

Of these examples the first six are determined through distributions of random vectors with a finite number of coordinates, while in the seventh one the Poisson polygon does not admit such a representation. However, it is still simple, since the corresponding distribution is determined by only one parameter (the intensity of the line process). More complicated examples appear as limit distributions for unions or convex hulls of random sets [Mol93a, Chapter 8]. Further models of random compact sets are discussed in [SS94, pp. 125–139].

To construct the *Boolean model* we take a Poisson point process Π_λ and a sequence of mutually independent random compact sets $\Xi_0, \Xi_1, \Xi_2, \ldots$ (for example, random balls), which are also independent of Π_λ. Then replace each point of $\Pi_\lambda = \{x_1, x_2, \ldots\}$ by the appropriately shifted corresponding set and take their union. The resulting union set Ξ is said to be the Boolean model. This procedure can be formulated mathematically as follows:

$$\Xi = \bigcup_{i:\, x_i \in \Pi_\lambda} (x_i + \Xi_i). \tag{2.3}$$

The points x_i are called *germs* and the random set Ξ_0 is said to be the 'typical' *grain* of the Boolean model Ξ. (Usually we omit the word 'typical'.) The set Ξ is sometimes called the Poisson *germ–grain model*. The general concept of germ–grain model includes also those cases when the point process of germs is not Poisson and allows the grains and the germs to be dependent. The value of λ is said to be the *intensity* of the Boolean model. To obtain a non-trivial union set the grain must satisfy additional conditions, for example, the second (or dth) moment of the radius of the circumscribed circle of Ξ_0 must be finite [Hei92b].

In the simplest case $\Xi_0 = \{o\}$ is the origin, and the Boolean model Ξ coincides with the point process Π_λ of germs.

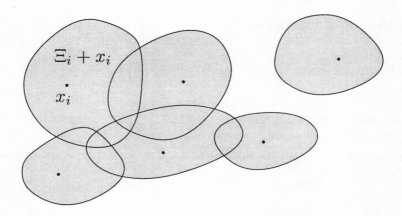

Figure 2.1 The union of 6 grains Ξ_i shifted by the points x_i, $1 \le i \le 6$. Note that these points x_i and the boundaries covered by other grains are invisible.

Notes to Section 2.3

Random closed sets. To give a formal definition of a random closed set, the space $\mathcal{F}$ of closed sets in $\mathbf{R}^d$ must be furnished with a σ-algebra. This σ-algebra, σ_f, is the minimum σ-algebra containing families of the type

$$\mathcal{F}_K = \{F \in \mathcal{F} : F \cap K \neq \emptyset\}$$

for K running through the family $\mathcal{K}$ of compact sets in $\mathbf{R}^d$. Now a random closed set is defined to be an $\mathcal{F}$-valued random element measurable with respect to σ_f [Mat75, SKM95, Mol93a]. This construction ensures measurability of many interesting geometric functionals of a random closed set. For example, the Lebesgue measure of a random closed set (if it is finite) is a random variable. Moreover, usual set-theoretic operations (those with closed values) preserve the measurability of their operands.

The distribution of a random closed set X is determined uniquely by the capacity functional

$$T_X(K) = \mathbf{P}\{X \cap K \neq \emptyset\}, \quad K \in \mathcal{K}.$$

This functional plays the same role as the distribution function of a random variable or the finite-dimensional distributions of a random function. It is possible to formulate conditions on a general functional $T : \mathcal{K} \mapsto [0, 1]$ such that it is the capacity functional of the (necessarily unique) random closed set, see [Mat75, Mol93a]. A random closed set is said to be *stationary* (*isotropic*) if its distribution is invariant with respect to all non-random translations (rotations). The capacity functional is then also invariant with respect to the same motions.

A random closed set X is said to be convex (compact) if almost all its realisations are convex (compact) sets. These geometrical assumptions make it possible to determine the distribution of X more effectively, see [Mol93a, SW86, Vit83].

Hausdorff metric. The space $\mathcal{K}$ of compact sets can be metrised by means of the so-called Hausdorff metric, which defines the distance between two compact sets K and K_1 by

$$\rho_H(K, K_1) = \inf \{r \geq 0 : K \subset K_1^r, \ K_1 \subset K^r\} , \qquad (2.4)$$

where

$$K^r = K \oplus B_r = \{x : \rho(x, K) \leq r\} \qquad (2.5)$$

is the set of all points with distance to K less than r. This metric is used to formulate properties of random sets and set-valued estimators.

Poisson polygon. Let us consider the Poisson point process on the space $(0, 2\pi] \times [0, \infty)$ with intensity measure $\frac{\lambda}{2\pi}\mu_2$, where $\mu_2 = $ A is the two-dimensional Lebesgue measure (area). Each point determines a line with normal given by the first coordinate and distance to the origin given by the second one. This Poisson point process corresponds to the family of lines (line network) called the Poisson line process of intensity λ. These lines divide the plane into convex polygons. When shifted in such a way that the centre of gravity is in the origin, these polygons are considered to be realisations of a random closed set X called the *Poisson polygon* [SS94, p.125]. The value of λ is the only parameter of the corresponding distribution.

Mean values and higher moments of perimeter and area of X are given in [Mat75, SKM95, SS94]. In particular,

$$\mathbf{E}A(X) = \frac{4}{\pi\lambda^2} , \quad \mathbf{E}U(X) = \frac{4}{\pi} , \qquad (2.6)$$

and

$$\operatorname{Var} A(X) = \frac{8\pi^2 - 16}{\pi^2\lambda^4} , \quad \operatorname{Var} U(X) = \frac{2\pi^2 - 8}{\lambda^2} . \qquad (2.7)$$

Another random polygon, the so-called Dirichlet polygon, is defined as the typical polygon formed by a Dirichlet (or Poisson–Voronoi) mosaic. Given a realisation of the Poisson point process this mosaic divides the plane into sets of points that have the given point of the Poisson point process as their nearest neighbour [SS94, p. 128].

Existence of the Boolean model. If the grain Ξ_0 is sufficiently large, then the union set (2.3) can be trivial, i.e. its closure can coincide with the whole plane. The necessary and sufficient condition for the existence of the Boolean model is

$$\mathbf{E}\mathsf{mes}(\Xi_0^r) < \infty \qquad (2.8)$$

for some $r > 0$. In turn, (2.8) follows from $\mathbf{E}\operatorname{diam}(\Xi_0)^d < \infty$, where $\operatorname{diam}(\Xi_0)$ is the diameter of the minimum circumscribed ball over Ξ_0. Indeed, $\mathsf{mes}(\Xi_0^r) <$

$(\text{diam}(\Xi_0)/2 + r)^d b_d$, where b_d is the volume of the unit ball in $\mathbf{R}^d$. Furthermore, $\text{diam}(\Xi_0) \leq 2\|\Xi_0\|$, where

$$\|\Xi_0\| = \sup\{\|x\|\colon x \in \Xi_0\}\,.$$

To avoid unnecessary complications we usually assume that $o \in \Xi_0$ almost surely. Then the Boolean model Ξ exists if $\mathbf{E}\|\Xi_0\|^d < \infty$.

Robbins' theorem. The Lebesgue measure of a random compact set X is a random variable. Its expectation can be computed as

$$\mathbf{Emes}(X) = \int_{\mathbf{R}^d} \mathbf{P}\{x \in X\}\, dx\,. \tag{2.9}$$

The expectation on the left-hand side exists as soon as the integral exists. This fact was first explicitly formulated and used by Robbins [Rob44, Rob45] in connection with applications to the bombing problem. In fact, (2.9) is a simple consequence of Fubini's theorem.

Similar results are valid for higher moments, for example,

$$\mathbf{E}[\mathsf{mes}(X)]^2 = \int_{\mathbf{R}^d}\int_{\mathbf{R}^d} \mathbf{P}\{\{x_1, x_2\} \subset X\}\, dx_1 dx_2\,. \tag{2.10}$$

Germ–grain model. Equivalently, the Boolean model can be constructed as a Poisson point process on the space of compact sets, see [Mat75, p. 61] and [Wei91, WW93]. The shifted sets $(x_i + \Xi_i)$ are understood then as 'points' of the point process in the space of closed sets. The corresponding intensity measure on $\mathcal{K}$ is defined to be the product of the Lebesgue measure on $\mathbf{R}^d$ and the probability measure on the space $\mathcal{K}_0$ of 'centred' compact sets, which is the distribution of the (centred) typical grain (for example, the sets from $\mathcal{K}_0$ may be chosen to have the centre of gravity at the origin). The first gives a random location to a 'point' (set) in the space $\mathcal{K}$, while the second determines the distribution of the grain's shape, see also [Wei87]. The Boolean model is, thereupon, a particular case of the general germ–grain model [Han81, Hei92b]. A general condition for the existence of germ–grain models was given in [Hei92b].

Decomposition theorem. Weil and Wieacker [WW87] established a decomposition theorem which states that rather general random closed sets can be obtained as germ–grain models. The random set X is said to belong to the *extended convex ring* if $X \cap W$ belongs to the *convex ring* for each convex compact set W. The latter means that $X \cap W$ is a union of no more than a finite number of convex compact sets. The family of finite unions of convex sets (convex ring) is denoted by $\mathcal{R}$.

Theorem 2.1 (see [WW87]) *If X is a random closed set with values in the extended convex ring, then there exists a point process $\Psi = \{Y_1, Y_2, \ldots\}$ on the family $\mathcal{C}$ of convex compact sets such that*

$$X = Y_1 \cup Y_2 \cup \cdots\,. \tag{2.11}$$

If the distribution of X is invariant with respect to a certain group, then the point process Ψ can be chosen to be invariant in distribution with respect to the same group.

PROOF is based on the measurable selection theorem, see [CV77]. This theorem provides existence of the decomposition [WW84], while its group-invariance is established in [WW87].   □

Let us associate with each Y_i from (2.11) a point $c(Y_i)$ in such a way that $c(Y_i + x) = Y_i + x$ for all x. For instance, $c(Y_i)$ can be the centre of gravity of Y_i. Then we get a representation of X which is similar to (2.3)

$$X = \bigcup_i (c(Y_i) + Y_i^0)$$

with $Y_i^0 = Y_i - c(Y_i)$. Note that, in contrast to (2.3), the associated points (or germs), $c(Y_i)$, and the centred sets (or grains), Y_i^0, can be dependent. The Boolean model appears then in the case when they are independent and the associated points form the Poisson point process.

Furthermore, any random closed set X can be represented as a union-set for some point process on $\mathcal{K}$, i.e. $X = Y_1 \cup Y_2 \cup \cdots$ for point process $\{Y_1, Y_2, \ldots\}$ on the space $\mathcal{K}$ of compact sets.

2.4   Examples of Boolean Models

We begin with the three simulated examples of the Boolean model shown in Figure 2.2. All these models have the same intensity $\lambda = 0.0062$ on the digitised computer screen. The samples are shown in windows of size 200×150 pixels. The left-hand figure is obtained for a grain equal to the convex hull of four points uniformly distributed in a square of $[-20, 20] \times [-20, 20]$ pixels. The grain of the second Boolean model is a random ellipse with axes parallel to the coordinate axes. The lengths of the axes are random and uniformly distributed in $[0, 20]$ and $[0, 14]$ (in pixels) respectively. The right-hand figure is constructed for a deterministic grain: namely a parallelogram with vertices $(0, 0)$, $(14, 0)$, $(30, 10)$ and $(16, 10)$. These simulated figures show that the Boolean model can have a considerable variety of shapes even for simple grains.

Over the course of time the Boolean model has been applied to various practical problems. Basic references are given in [Hal88, Ser82, Sto79, SKM95]. It is intuitively clear that the Boolean model can describe spatial patterns containing independent randomly located overlapping components. The list of applications includes bombing (germs are points of impact and grains are damaged regions caused by each separate bomb) [Ahu78, NS72, Rob44,

Figure 2.2   Simulated examples of the Boolean model.

Rob45], the microstructure of paper (grains are elementary fibres which can be described as random rectangles) [Dod71, MSF93], tumour growth [CH92], spatial patterns of heather in the countryside [Dig81], geological deposits [Ber93, Che95, OS80, Ter94], crystallisation in metals and crystal growth [MC95, Gil62, Kol37], the microstructure of dough [BS91], patterns in photographic emulsions [Lya86, Lya88], structure of materials [BMF86, OT88, QCCJ92], systems of water droplets and many others. Some of these papers did not use the strict definition of the Boolean model but used the same idea to construct relevant spatial patterns.

Usual candidates for the typical grain are:

1. Random ball.
2. Random ellipse.
3. Random rectangle.
4. Deterministic set.
5. Random polygon, e.g., Poisson polygon or Dirichlet polygon, see [SS94, pp. 125, 128].
6. Randomly rotated and/or scaled deterministic set.
7. Convex-stable or union-stable sets, see [Mol93a, Chapter 8].

Figure 2.3 shows a real picture, which (now at least visually) can be considered to be a Boolean model with polygonal grains. This photograph (provided by J.-L. Quenec'h) represents a planar section of WC-Co alloy structure. From a first glance it is reasonable to suppose that the black component is a Boolean model with a polygonal grain. Similar images were studied in [QCCJ92].

Another example (provided by D. Stoyan) is shown in Figure 2.4. It seems plausible that the black set can be interpreted as a Boolean model with nearly circular grains.

Figure 2.3 Planar pattern obtained as union of crystals (reproduced by permission of J.-L. Quenec'h).

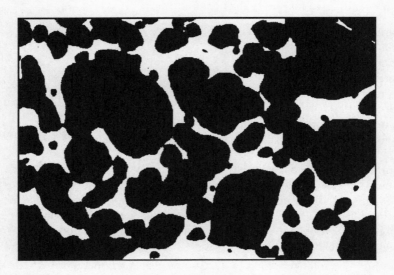

Figure 2.4 Pores in white bread (reproduced by permission of D. Stoyan).

Notes to Section 2.4

Boolean models in stochastics. The Boolean model is used in random set theory and spatial statistics to model stationary sets, see [Cre91, Ser82, SKM95] and references therein. The corresponding problems there are related to estimation techniques and hypothesis testing. This is also the subject of the present book.

At the same time, the Boolean model serves as the basic model in continuum percolation theory, see [Hal85b, Hal88, Pen91, MR94]. The main problem is the existence of unbounded connected covered (or uncovered) sets.

Another application of the Boolean model is related to coverage problems, see [Eva90, Hal85a, Hal85c, HG88, Nau79, Sta89]. The principal aim is to find the probability of coverage of a fixed set K by Ξ. Because of certain biological applications one often considers the coverage problem for the Boolean model on the circle or sphere [Hüs82, KM63, Mil69, She72, SH82]. Another family of coverage problems is related to the study of the time of total coverage for time-dependent Boolean models and Johnson–Mehl tessellations [Chi95, CCH95, VS88]. The complement to the Boolean model (or the uncovered region) is of most interest in the theory of coverage [Ald89, Hal88]. Under certain conditions and the isotropy assumption Hall [Hal85a, Hal88] proved that the typical uncovered region is distributed like a polyhedron bounded by a stationary and isotropic Poisson net of hyperplanes. He also found the asymptotic probability of total coverage for the Boolean model with high intensity and small grain. Further results of this kind can be found in [Mol96a].

2.5   Characteristics and Important Parameters

The Boolean model defined above is a *stationary* (or homogeneous) random closed set. This means that the distribution of Ξ is the same as the distribution of $\Xi + x$ for every x. If the grain Ξ_0 is *isotropic* (i.e. the distribution of Ξ_0 is rotation-invariant), then Ξ is also isotropic (Ξ can be isotropic also for non-isotropic Ξ_0, see p. 24 and [MS94a]).

The *ergodicity property* of the Boolean model allows the use of spatial averages to replace probability averages (or expectations). For example, the *area fraction* (part of the area covered) is equal to the probability that a given point is covered by Ξ. The area fraction is denoted by p or A_A. It is known that

$$p = 1 - \exp\{-\lambda \mathbf{E} A(\Xi_0)\}, \qquad (2.12)$$

where $A(\Xi_0)$ is the area of the typical grain. In spaces of higher dimensions p is called the *volume fraction*.

Further results for *specific values* of geometric functionals are considered in [WW84, WW93]. These values are defined as spatial densities of the relevant geometric functionals (volume, boundary length, surface area, etc.), see p. 27. For example, the *specific boundary length* is given by

$$L_A = \lambda(1 - p)\mathbf{E}\mathsf{U}(\Xi_0)\,, \tag{2.13}$$

where $\mathsf{U}(\Xi_0)$ is the perimeter of Ξ_0. Roughly speaking, the value L_A is equal to the expected boundary length of Ξ in a window of unit area.

The *specific connectivity number* χ_A (or specific Euler–Poincaré characteristic) can be viewed as the expected difference between the number of clumps and the number of holes per unit area, see Section 3.4. If Ξ_0 is convex and *isotropic*, then

$$\chi_A = (1 - p)\left(\lambda - \frac{\lambda^2}{4\pi}(\mathbf{E}\mathsf{U}(\Xi_0))^2\right)\,, \tag{2.14}$$

see [Mil76, Wei88, WW84]. Note that if Ξ contains only disjoint convex particles, then χ_A simply equals the number of particles per unit area. For a general almost surely convex grain Ξ_0 the specific connectivity number of Ξ can be expressed through the Aumann expectation of the grain Ξ_0, see [Sch92, Wei95].

Among several definitions of expectations of random sets (see [SS94]) the Aumann expectation is the best adjusted for convex random sets. It characterises the mean shape of a set and is informative if the grain is anisotropic. This expectation can be introduced as follows. First, let us note that each *convex* set F is determined uniquely by its support function:

$$h(F, u) = \sup\{\langle u, x\rangle\colon x \in F\}\,, \tag{2.15}$$

where u is a unit vector and

$$\langle u, x\rangle = \sum_{i=1}^{d} u_i x_i$$

is the scalar product of u and x. The support function equals the maximum distance between the origin and the line (or plane in higher dimensions) orthogonal to u and moving in the direction of u, see Figure 2.5. By convexity, F can be retrieved as intersection of half-planes determined by its support function.

If Ξ_0 is a random compact set, then $h(\Xi_0, u)$ is a random variable. For the random set Ξ_0 its expected support function (if it exists) is again the support function of a convex set called the Aumann expectation of Ξ_0, see [SS94, pp. 109–110] and [Vit88]. In other words, the *Aumann expectation*, $\mathbf{E}\Xi_0$, of Ξ_0 satisfies the following identity:

$$h(\mathbf{E}\Xi_0, u) = \mathbf{E}h(\Xi_0, u)\,. \tag{2.16}$$

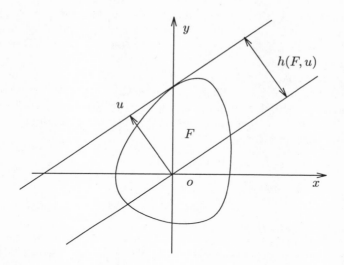

Figure 2.5   Support function of F.

This means that the support function of $\mathbf{E}\Xi_0$ is equal to the expectation of the support function of Ξ_0. Therefore, the Aumann expectation is always convex even when taken for non-convex (even non-random) sets. If Ξ_0 is isotropic, then $\mathbf{E}\Xi_0$ is a ball with diameter equal to the expected mean width of Ξ_0 (on the plane this yields equality between the expected perimeter of Ξ_0 and the perimeter of $\mathbf{E}\Xi_0$).

The area fraction of a stationary random set Ξ satisfies $p = \mathbf{P}\left\{o \in \Xi\right\}$. This suggests considering also two-point covering probabilities:

$$C(v) = \mathbf{P}\left\{\{o, v\} \subset \Xi\right\}. \tag{2.17}$$

This function is said to be the *covariance* of Ξ. The covariance is simply related to the covariance function of the indicator random field

$$\zeta(v) = \mathbf{1}_{v \in \Xi} = \begin{cases} 1, & v \in \Xi, \\ 0, & \text{otherwise} \end{cases}$$

generated by Ξ, since

$$C(v) = \mathbf{E}[\zeta(o)\zeta(v)].$$

In the isotropic case $C(v)$ depends only on the length, $\|v\|$. Further properties of the covariance can be found in [SKM95, Chapters 3, 6]. The function

$$q(v) = 1 + \frac{C(v) - p^2}{(1 - p)^2} = \exp\{\lambda\mathbf{E}\mathbf{A}(\Xi_0 \cap (\Xi_0 - v))\} \tag{2.18}$$

will also be of use. Again, the area must be replaced by the volume for higher dimensions. Clearly, $q(o) = (1 - p)^{-1}$, and $q(v) \to 1$ if $\|v\| \to \infty$. If Ξ_0 is convex, then $q(tv)$, $t \geq 0$, is a decreasing function for each $v \in \mathbf{R}^d$ (note that the converse is not true!).

Since the area $A(\Xi_0 \cap (\Xi_0 - v))$ (its expectation is said to be the set-covariance function of Ξ_0 [Ser82, p. 272] and [SS94, p. 122]) is complicated to evaluate even for simple grains, it is not easy to work with the covariance. Sometimes in the isotropic case it is advisable to approximate it by the *exponential* covariance formula

$$C_e(v) = p(1 - p)e^{-\alpha \|v\|} + p^2, \quad r \geq 0, \qquad (2.19)$$

where

$$\alpha = L_A / (\pi p(1 - p)), \qquad (2.20)$$

with L_A given by (2.13), see [SKM95, p. 205].

The area fraction is equal to the probability that a point (say, the origin) is covered by Ξ. The covariance is given through two-point covering probabilities. As an extension, it is possible to consider the probability that Ξ hits a fixed compact set K. This probability is denoted by $T_\Xi(K)$ and is known as the *capacity* (or hitting) *functional* of Ξ. It is given by

$$T_\Xi(K) = \mathbf{P}\{\Xi \cap K \neq \emptyset\} = 1 - \exp\{-\lambda \mathbf{E} A(\Xi_0 \oplus \check{K})\}, \qquad (2.21)$$

where $\oplus$ is the Minkowski addition and $\check{K} = \{-y : y \in K\}$ is the reflection of K with respect to the origin, i.e.

$$\Xi_0 \oplus \check{K} = \{x - y : x \in \Xi_0, y \in K\}.$$

In particular, if K is a disk of radius r centred at the origin, then $\Xi_0 \oplus \check{K}$ is the set of all points at distance no more than r from Ξ_0 (this set is denoted by Ξ_0^r and is said to be the r-envelope of Ξ_0), see Figure 2.6. Formula (2.21) is valid in higher-dimensional spaces provided that the area is replaced by the Lebesgue measure $\mathrm{mes}(\cdot)$ (volume for $d = 3$).

The value $T_\Xi(K)$ equals the area fraction of the set $\Xi \oplus \check{K}$. Indeed,

$$\mathbf{P}\{x \in \Xi \oplus \check{K}\} = \mathbf{P}\{o \in \Xi \oplus \check{K}\} = \mathbf{P}\{K \cap \Xi \neq \emptyset\}$$

for all $x \in \mathbf{R}^2$.

Note that the exact probability that Ξ *covers* a compact set is usually impossible to find theoretically. It is related to difficult coverage problems, see [Hal88].

Notes to Section 2.5

Capacity functional of the Boolean model. Formula (2.21) for the capacity functional of the Boolean model can be derived using a number of different methods.

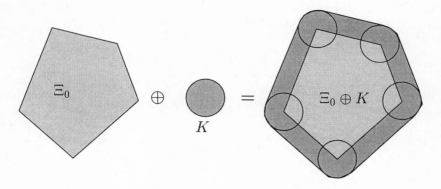

Figure 2.6   Minkowski addition of two sets.

We will follow the proof in [SKM95, p. 65] for a general (non-stationary) Poisson process of germs, Π_Λ, with locally finite intensity measure Λ.

Let Π'_Λ be the set of all points $x_n \in \Pi_\Lambda$ with $(\Xi_n + x_n) \cap K \neq \emptyset$. In other words, Π'_Λ is a result of a thinning procedure applied to the Poisson point process Π_Λ. Then each point $x \in \Pi_\Lambda$ belongs also to Π'_Λ with probability

$$p(x) = \mathbf{P}\left\{(\Xi_0 + x) \cap K \neq \emptyset\right\} = \mathbf{P}\left\{x \in \check{\Xi}_0 \oplus K\right\}.$$

Then, by the Poisson property and Fubini's theorem,

$$
\begin{aligned}
T_\Xi(K) = \mathbf{P}\left\{\Xi \cap K \neq \emptyset\right\} &= 1 - \mathbf{P}\left\{\Pi'_\Lambda = \emptyset\right\} \\
&= 1 - \exp\left\{-\int_{\mathbf{R}^d} p(x)\Lambda(dx)\right\} \\
&= 1 - \exp\{-\mathbf{E}\Lambda(\check{\Xi}_0 \oplus K)\}.
\end{aligned}
$$

In the stationary case, $\Lambda = \lambda\mathrm{mes}$, whence $\Lambda(\check{\Xi}_0 \oplus K) = \lambda\mathrm{mes}(\Xi_0 \oplus \check{K})$.

Formula (2.12) for the covering probability easily follows from (2.21), since, by stationarity,

$$p = \mathbf{P}\left\{x \in \Xi\right\} = \mathbf{P}\left\{o \in \Xi\right\} = T_\Xi(\{o\}) = 1 - \exp\{-\lambda\mathbf{E}\mathrm{mes}(\Xi_0)\}.$$

By Robbins' theorem (see (2.9)),

$$\mathbf{E}\mathrm{mes}(\Xi \cap W) = \int_W \mathbf{P}\left\{x \in \Xi\right\} dx = p\,\mathrm{mes}(W).$$

Thus, p is equal to the expected part of the volume covered by the Boolean model Ξ in a window W. Formula (2.18) for the covariance also follows from (2.21), since

$$C(v) = 2p - T(\{o, v\}).$$

Non-Poisson point processes of germs. Suppose for the moment that the point process of germs, Ψ, is not Poisson, but the germs and grains are independent. Then

$$1 - T_\Xi(K) = \mathbf{E}\left[\prod_{x_i \in \Psi}\left(1 - \mathbf{1}_{K \oplus \check{\Xi}_i}(x_i)\right)\right]$$

$$= \mathbf{E}\left[\prod_{x_i \in \Psi}\mathbf{E}\left[\left(1 - \mathbf{1}_{K \oplus \check{\Xi}_i}(x_i)\right) | \Psi\right]\right]$$

$$= \mathbf{E}\left[\prod_{x_i \in \Psi}\mathbf{P}\left\{x_i \notin K \oplus \check{\Xi}_0\right\}\right].$$

Note that the probability generating functional of Ψ is defined by

$$G[f(\cdot)] = \mathbf{E}\left[\prod_{x_i \in \Psi}f(x_i)\right],$$

see [DVJ88]. Here $f(\cdot)$ is a function such that $|\log f(x)|$ is integrable over $\mathbf{R}^d$. The Poisson assumption implies

$$G[f(\cdot)] = \exp\left\{-\lambda\int_{\mathbf{R}^d}(1 - f(x))dx\right\}.$$

Thus, for the general germ process Ψ,

$$T_\Xi(K) = 1 - G\left[\mathbf{P}\left\{(\cdot) \notin \check{\Xi}_0 \oplus K\right\}\right].$$

Boolean models and stationary random fields. The indicator function $\zeta(x) = \mathbf{1}_{x \in \Xi}$, $x \in \mathbf{R}^d$, of the Boolean model (as well as of any stationary random set) is a stationary random field. Its mean characteristics are related to those of the corresponding Boolean model. For example,

$$\mathbf{E}\zeta(x) = \mathbf{P}\left\{\zeta(x) = 1\right\} = p,$$
$$\mathbf{E}[\zeta(x)\zeta(y)] = \mathbf{P}\left\{\{x, y\} \subset \Xi\right\} = C(y - x).$$

It is, however, not clear how spectral characteristics of ζ are related to geometric properties of the Boolean model. It should be noted that, for the typical grain of a.s. zero Lebesgue measure, the function $\zeta(x)$ is a.s. equal to 0 for all $x \in \mathbf{R}^d$. Otherwise $\zeta(x)$ is non-trivial.

Uniqueness of parameters. We have already seen that the intensity of the germs and the distribution of the typical grain determine the distribution of a Boolean model. The natural question as to whether two Boolean models with different parameters can share the same distribution was solved in [PS87], see also [PS88, Rat95, Sch91]. It was proved that two Boolean models with the same distribution have equal intensities and their grains are equal in distribution up to a shift. Below is an outline of the proof following [PS87].

First, one establishes that the map $K \mapsto c(K)$ with $c(K)$ being the centre of the circumscribed circle (ball) of K is measurable and, moreover, continuous with respect to the Hausdorff metric on the family of compact sets. Furthermore, for any probability measure P on σ_f,

$$\theta(\mathcal{A}) = \int_{\mathbf{R}^d} P(\mathcal{A} - x) dx, \quad \mathcal{A} \in \sigma_f, \tag{2.22}$$

defines a measure on σ_f, where $\mathcal{A} - x = \{F - x : F \in \mathcal{A}\}$. If P is the distribution of Ξ_0, then, by Robbins' theorem,

$$\mathbf{E}\mathrm{mes}(\Xi_0 \oplus \check{K}) = \int_{\mathbf{R}^d} P(\mathcal{F}_{K-x}) dx = \theta(\mathcal{F}_K).$$

Let us define

$$\mathfrak{t}(\mathcal{A}) = \bigcup_{x \in \mathbf{R}^d} (\mathcal{A} \cap c^{-1}(o)) + x,$$

i.e. $\mathfrak{t}(\mathcal{A})$ consists of all translates of sets $F \in \mathcal{A}$ with corresponding centre $c(F) = o$. Clearly, $\mathfrak{t}(\mathcal{A}) \in \sigma_f$ as soon as $\mathcal{A} \in \sigma_f$. If Ξ_0 is the grain of the Boolean model, then it is possible to shift it in such a way that the probability distribution P of the shifted set satisfies

$$P(\mathcal{A}) = P(\mathfrak{t}(\mathcal{A})), \quad \mathcal{A} \in \sigma_f. \tag{2.23}$$

We can restrict ourselves to distributions satisfying (2.23). Their equality corresponds to equality of the corresponding random sets up to a translation.

Consider now two Boolean models with intensities λ_1 and λ_2 and distributions P_1 and P_2 for their centred grains. Define the corresponding measures θ_1 and θ_2 by (2.22). For each Borel set $H \subset \mathbf{R}^d$,

$$\theta_1(\mathcal{A} \cap c^{-1}(H)) = \int_{\mathbf{R}^d} P_1((\mathcal{A} - x) \cap [c^{-1}(H - x) \cap c^{-1}(\{o\})]) dx,$$

since P_1 is concentrated on $c^{-1}(\{o\})$. Furthermore, $c^{-1}(H - x) \cap c^{-1}(\{o\})$ equals $c^{-1}(\{o\})$ if $x \in H$ and is empty otherwise. Thus,

$$\theta_1(\mathcal{A} \cap c^{-1}(H)) = \int_H P_1(\mathcal{A} - x) dx,$$

whence

$$\lambda_1 P_1(\mathcal{A} - x) = \lambda_2 P_2(\mathcal{A} - x), \quad \mathcal{A} \in \sigma_f,$$

for almost all $x \in \mathbf{R}^d$. By (2.23),

$$\begin{aligned} \lambda_1 P_1(\mathcal{A}) = \lambda_1 P_1(\mathfrak{t}(\mathcal{A})) &= \lambda_1 P_1(\mathfrak{t}(\mathcal{A} - x)) \\ &= \lambda_1 P_1(\mathcal{A} - x) \\ &= \lambda_2 P_2(\mathcal{A} - x) = \cdots = \lambda_2 P_2(\mathcal{A}). \end{aligned}$$

for all $\mathcal{A} \in \sigma_f$. Letting $\mathcal{A} = \mathcal{F}$ yields $\lambda_1 = \lambda_2$, whence also $P_1 = P_2$. □

Isotropic Boolean models. Isotropy of the grain Ξ_0 yields isotropy of the corresponding Boolean model Ξ. However, the reverse (plausible at first glance) statement is not true. Namely, an isotropic Boolean model can have an anisotropic typical grain [MS94a]. An example is easy to construct by using the fact that a Boolean model with grain $\Xi_0 + x$ has the same distribution as a Boolean model with grain Ξ_0. Thus, it is always possible to shift Ξ_0 to get an anisotropic random set, which (taken as a grain) generates the same Boolean model. However, this anisotropy of the grain cannot 'range very far'.

Theorem 2.2 (see [MS94a]) *If Ξ is an isotropic Boolean model with typical grain Ξ_0, then there exists a random vector $\zeta(\Xi_0)$, such that $\Xi_0 - \zeta(\Xi_0)$ is an isotropic random set. For instance, $\zeta(\Xi_0)$ can be chosen to be the centre of the circumscribed ball of Ξ_0.*

PROOF. By the uniqueness property of the Boolean model, $\omega\Xi_0$ coincides in distribution with $\Xi_0 + \xi_\omega$ for any rotation ω and corresponding shift ξ_ω. Furthermore, $\omega\zeta(\Xi_0)$ coincides in distribution with $\zeta(\Xi_0) + \xi_\omega$ for $\zeta(\Xi_0)$ being equal to the centre of the minimum circumscribed ball of Ξ_0. After subtraction of these two equations we conclude that $\omega(\Xi_0 - \zeta(\Xi_0))$ has the same distribution as $\Xi_0 - \zeta(\Xi_0)$, which entails the isotropy of the shifted grain. □

Aumann expectation, Minkowski addition. The Aumann expectation is only one possible definition for expectation of a random set, see [Sto89, SS94]. However, it provides definitely better and more tractable results for convex random sets. The definition of the Aumann expectation given above (see p. 19) uses the isometry between the space of convex sets and the family of support functions (sublinear functions on the unit sphere).

Let us consider another definition (see [AV75, Aum65] and [SS94, p. 109]), which yields the same result if the basic probability space contains no atoms [Vit91]. Let Ξ_0 be a random set with $\mathbf{E}\|\Xi_0\| < \infty$. A random point $\xi \in \mathbf{R}^d$ is said to be a *selection* of Ξ_0 if $\xi \in \Xi_0$ with probability one. Then the Aumann expectation of Ξ_0 is defined by

$$\mathbf{E}\Xi_0 = \{\mathbf{E}\xi : \ \xi \text{ is a selection of } \Xi_0, \ \mathbf{E}\xi \text{ exists}\}.$$

The Aumann expectation appears in the strong law of large numbers for the Minkowski addition of random compact sets. The Minkowski addition (see Figure 2.6) is defined element-wise as

$$F_1 \oplus F_2 = \{x_1 + x_2 : \ x_1 \in F_1, \ x_2 \in F_2\}.$$

The strong law of large numbers states that $n^{-1}(X_1 \oplus \cdots \oplus X_n)$ converges almost surely in the Hausdorff metric to $\mathbf{E}X_1$ for any sequence of independent identically distributed random compact sets $X_1, X_2, \ldots$ with $\mathbf{E}\|X_1\| < \infty$, see [AV75]. Further results in this direction include the central limit theorem, the law of the iterated logarithm and the elementary renewal theorem, see [Cre79, GHZ83, Lya82, MOK95, Wei82] etc.

2.6   Individual and Aggregate Parameters

All parameters of the Boolean model can be classified according to the following scheme. One group includes *aggregate* (or macroscopic) parameters which characterise the set Ξ 'in the whole'. Examples are p, L_A and $C(v)$. They are *directly* observable and can be estimated as spatial averages from an observation of Ξ. The ergodicity property ensures that the corresponding estimators converge with probability one to the theoretical values if the window of observations expands to the whole plane. The capacity functional $T_\Xi(K)$, $K \in \mathcal{K}$, of the Boolean model is also an aggregate parameter that determines uniquely the distribution of Ξ.

Individual (or microscopic) parameters form the other group of parameters. Typical examples are the expected area, $\mathbf{EA}(\Xi_0)$, and expected perimeter, $\mathbf{EU}(\Xi_0)$, of the grain, the distributions of $A(\Xi_0)$ and $U(\Xi_0)$, the Aumann expectation, $\mathbf{E}\Xi_0$, and the distribution of Ξ_0 as a random compact set given by the capacity functional $T_{\Xi_0}(K)$. The intensity λ can be considered an individual parameter, that of the Poisson point process of germs. The individual parameters are not directly observable. However, they are of the most interest, such as when fitting a proper model to real data. Only knowledge of the individual parameters makes it possible to simulate the Boolean model, and then to use tests based on simulations. Roughly speaking, aggregate parameters describe the visual picture, while the individual parameters reveal the picture's nature.

As individual parameters are not observable, they can be estimated only through estimates of aggregate parameters, as soon as relationships between their theoretical counterparts are found. It should be noted that sometimes individual parameters can be directly observable. For example, if the grain Ξ_0 is a curve, then all shifted grains can be traced. If the grain is a segment (the corresponding Boolean model is said to be the Boolean segment process), then it is possible to determine the orientation distribution of the segment by means of a standard technique from the theory of fibre processes, see [SKM95]. For vast samples of Ξ it is possible to meet separate intact grains with no overlappings. However, all methods based on this possibility require hard manual work, and also the probability of meeting separate grains decreases rapidly with the growth of the area fraction, see [SKM95, p. 73]. We will consider neither this simplest case of observable grains nor methods related to the manual analysis of the union set.

In this book we will assume (unless otherwise stated) that the grain Ξ_0 is *convex* and *regular closed* (coincides with the closure of its interior) with probability one.

Mathematical treatments of statistics of the Boolean model are presented in [Cre91, CL87, Hal88, Ser82, SKM95]. Recent surveys can be found in [AFM90, Mol91c, Mol92, Mol95, SPRB94, Sch92, Wei95].

Notes to Section 2.6

Ergodicity. The ergodic property of the Boolean model was studied in [Hei92b]. As long as the existence condition (2.8) is valid, the Boolean model Ξ is mixing and ergodic, i.e. for all $\mathcal{A}_1, \mathcal{A}_2 \in \sigma_f$

$$\mathbf{P}((\mathcal{A}_1 + x) \cap \mathcal{A}_2) \to \mathbf{P}(\mathcal{A}_1)\mathbf{P}(\mathcal{A}_2) \quad \text{as} \quad \|x\| \to \infty, \qquad (2.24)$$

and

$$\frac{1}{\text{mes}(W)} \int\limits_{W} \mathbf{P}((\mathcal{A}_1 + x) \cap \mathcal{A}_2)dx \to \mathbf{P}(\mathcal{A}_1)\mathbf{P}(\mathcal{A}_2) \quad \text{as} \quad W \uparrow \mathbf{R}^d, \qquad (2.25)$$

where $\mathbf{P}$ is the distribution of Ξ. From this it follows in particular that spatial averages of geometric functionals converge to the corresponding expected values.

Intrinsic volumes. The *intrinsic volumes* $V_j(K)$, $0 \le j \le d$, of a convex compact set K can be defined by means of the following *Steiner formula*:

$$\text{mes}(K^r) = \sum_{j=0}^{d} r^{d-j} b_{d-j} V_j(K), \qquad (2.26)$$

where K^r has been defined by (2.5) and b_{d-j} is the volume of the unit ball in the space $\mathbf{R}^{d-j}$, see [Sch93b, p. 210] and [SW92, pp. 38–39]. In other words, $\text{mes}(K^r)$ is a polynomial of degree d with respect to r, so that the corresponding coefficients are determined uniquely. Then $V_d(K) = \text{mes}(K)$ is the volume or d-dimensional Lebesgue measure, the value $V_{d-1}(K)$ is half of the surface area of K, and $V_1(K)$ is proportional to the mean width of K, more precisely,

$$V_1(K) = \frac{1}{2b_{d-1}} \int\limits_{\mathbf{S}^{d-1}} h(K, u)\mu_{d-1}(du).$$

Here $\mathbf{S}^{d-1}$ is the unit sphere in $\mathbf{R}^d$ and μ_{d-1} is the $(d-1)$-dimensional Hausdorff measure [SS94] (or surface area) on $\mathbf{S}^{d-1}$. The functional $V_0(K)$ is equal to 1 if K is non-empty and to 0 otherwise. The other functionals $V_{d-2}(K), \ldots, V_2(K)$ can be found as integrals of curvature functions (if K is smooth).

Remember that the family of finite unions of convex sets is called the *convex ring* and denoted by $\mathcal{R}$. The intrinsic volumes admit an *additive extension* for compact sets from the convex ring, see also [Sch93b, SW93, Wei83]. If K is the union of convex sets $K_1, \ldots, K_n$, then

$$V_j(K) = \sum_{i} V_j(K_i) - \sum_{i_1 < i_2} V_j(K_{i_1} \cap K_{i_2}) + \cdots + (-1)^{n+1} V_j(K_1 \cap \cdots \cap K_n)$$

is defined by the usual inclusion–exclusion formula. Then V_d and V_{d-1} retain their geometrical meanings and remain positive, while other functionals can become negative. The functional $V_0(K)$ is especially important. It is called the Euler–Poincaré characteristic of K and is usually denoted by $\chi(K)$.

The *positive extension* of the intrinsic volumes onto the convex ring can be defined as follows [Wei83]. For $x \in \mathbf{R}^d$ and K from the convex ring we will call a point $y \in K$ a metric projection of x onto K if there is a neighbourhood U of y such that $\|x - z\| > \|x - y\|$ for all $z \in U \cap K$, $z \neq y$, see Figure 2.7.

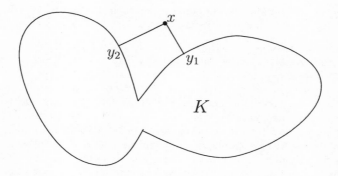

Figure 2.7 Two metric projections of x onto K.

For $r > 0$ let $\bar{c}_r(K, x)$ be the number of projections y of x with $\|y - x\| \leq r$. Then the expansion

$$\int_{\mathbf{R}^d} \bar{c}_r(K, x)dx = \sum_{j=0}^{d} r^{d-j}b_{d-j}\bar{V}_j(K) \tag{2.27}$$

defines the positive extension of the intrinsic volumes. If K is convex, then the integral on the left-hand side equals the volume on the left-hand side of the Steiner formula (2.26), whence $\bar{V}_j = V_j$ for convex compact sets.

Densities of the intrinsic volumes. If Ξ is a Boolean model with an a.s. convex (or even a.s. belonging to the convex ring $\mathcal{R}$) grain Ξ_0, then almost all realisations of Ξ belong to the *extended convex ring*, i.e. $\Xi \cap W \in \mathcal{R}$ a.s. for each compact convex set W.

Thus, it is possible to determine the values of (say, additively) extended intrinsic volumes, $V_j(\Xi \cap W)$, which become aggregate parameters of the Boolean model. The strong law of large numbers for intrinsic volumes was established in [WW84], see also [NZ79] for a general concept. In particular, the ergodic property of the Boolean model implies existence of the limit

$$D_j = \lim_{W \uparrow \mathbf{R}^d} \frac{V_j(\Xi \cap W)}{\mathrm{mes}(W)}. \tag{2.28}$$

For general j, the limit (2.28) can be found through multiple integrals with respect to the distribution of the grain [Wei88, Wei90, WW84]. Simpler results are obtained for the isotropic case and for $j = d$ or $j = d - 1$. For example, $D_d = p = 1 - \exp\{-\lambda \mathbf{E} \mathrm{mes}(\Xi_0)\}$ is the volume fraction, and

$$D_{d-1} = \mathbf{E} V_{d-1}(\Xi_0)(1 - p)$$

is the so-called surface area density of Ξ. Formulae of this kind are very important in stereology [Dav76, Dav78, DM77] and for parameter estimation of the Boolean model.

If ϕ is a general functional on the convex ring, then the limit of $\phi(\Xi \cap W)/\mathrm{mes}(W)$ as $W \uparrow \mathbf{R}^d$ (if it exists) is said to be the *spatial density* of ϕ.

Neyman–Scott processes. If the grain Ξ_0 is finite, then the corresponding Boolean model is known under the name of Neyman–Scott process, see [DVJ88, p. 245] and [SKM95, pp. 157–162]. The germs are called the parent points, and the points of the shifted grains (or clusters) $\Xi_n + x_n$ are called the daughter points. The parent points are not directly observable. Such a model was used in [NS72] to describe the locations of galaxies in space and the geometry of bombing. Other applications and references can be found in [SKM95, p. 162]. Some statistical problems are considered in [Mol91b]. It should be noted that consistent statistical procedures for Neyman–Scott processes are possible only under some assumptions on the distribution of daughter points in each cluster. Otherwise, the whole set $\Xi \cap W$ can be considered as a single cluster with many daughter points.

Boolean fibre processes. If the grain Ξ_0 is a random curve, then the Boolean model Ξ is called the *fibre process* (or surface process if Ξ_0 is a piece of a surface). For such Boolean models the orientation distribution of curves is of greatest interest. In general, such orientation characteristics can be estimated without referring to the Poisson property of the germ process, see [WW93, Section 6] and [SKM95, Chapter 9]. Statistical methods are usually based on the study of point processes resulting from the intersection of Ξ and a moving hyperplane or another 'testing system' of fibres, see [BCO94, CB94].

3

Estimation of Aggregate Parameters

In general, the aggregate parameters are of three types:

- numerical parameters;
- set-valued parameters;
- functional parameters.

In this chapter the first two groups are considered.

3.1 Area Fraction

The simplest parameter of the Boolean model is its area fraction, denoted by p or A_A. This is the part of the area covered by Ξ. Relevant estimation techniques were designed long ago, see the references in [Wei80, Chapter 3]. As a rule they have little to do with the Boolean model assumption and can be applied to every ergodic random set [Bad80, Hei92b, NZ79].

Similarly to Section 2.2, we consider a window W belonging to a family of convex expanding windows $W_s, s > 0$, so that $W \uparrow \mathbf{R}^2$ means $s \uparrow \infty$. Then an estimator of p is given by the area fraction measurement inside W:

$$\hat{p}_W = \frac{\mathsf{A}(\Xi \cap W)}{\mathsf{A}(W)}. \qquad (3.1)$$

If the set Ξ is ergodic (for the planar Boolean model the existence of the second moment of the diameter of the circumscribed circle of the grain is sufficient, and we always assume this), then $\hat{p}_W$ converges to p with probability one ($\hat{p}_W \to p$ a.s.) as $W \uparrow \mathbf{R}^2$. It means that the estimator given by (3.1) is *strong consistent*. Another possibility is to apply the point counting principle [Wei80, p. 64], i.e. to put

$$\hat{p}_{W,0} = \frac{\mathsf{N}(\Xi \cap W \cap \mathbb{Z}^2)}{\mathsf{N}(W \cap \mathbb{Z}^2)},$$

where $\mathbb{Z}^2$ is a lattice (grid) in the plane and $\mathsf{N}(\cdot)$ denotes the number of points belonging to the corresponding set. It is equivalently possible to evaluate $\hat{p}_W$ as the length fraction of one-dimensional sections of Ξ [Wei80, p. 61], i.e. the covered length divided by the total length of a system of lines inside W. All this can be summarised by the following stereological identity:

$$A_A = L_L = P_P.$$

These notations will become clearer after other two examples: L_A is the length of the line system per unit area, P_A is the number of points in the unit area, see also [SKM95, Chapter 11].

All corresponding estimators are strong consistent. Their asymptotic properties were studied in [Bad80, Mas82]. For example, $A(W)^{1/2}(\hat{p}_W - p)$ converges in distribution as $W \uparrow \mathbf{R}^2$ to a Gaussian random variable with zero mean and the variance

$$\sigma^2 = \int_{\mathbf{R}^2} (C(v) - p^2)dv < \infty \tag{3.2}$$

provided the latter integral is finite. In turn, $\sigma^2 < \infty$ if $\mathbf{E}A(\Xi_0)^2 < \infty$, i.e. the area (or volume) of the grain must have finite second moment. However, it is not easy to build a confidence interval for the area fraction. For this, it is necessary first to estimate the covariance function as is suggested either in [Mas82] or in Section 4.2.

The exact variance of $\hat{p}_W$,

$$\mathrm{Var}\,\hat{p}_W = A(W)^{-1} \int_{\mathbf{R}^2} \gamma_W(v)(C(v) - p^2)dv, \tag{3.3}$$

depends on the window through its *set-covariance function*

$$\gamma_W(v) = A(W \cap (W - v)).$$

For the exponential covariance (2.19) the value σ^2 from (3.2) can be expressed through simple numerical parameters of the sample, as

$$\sigma^2 = \int_{\mathbf{R}^2} p(1-p)e^{-\alpha\|v\|}dv = 2\pi p(1-p)\int_0^\infty re^{-\alpha r}dr = 2\pi p(1-p)\alpha^{-2}.$$

By substituting α from (2.20) we get

$$\sigma^2 = 2\pi^3 p^3 (1-p)^3 / L_A^2 \,. \tag{3.4}$$

Notes to Section 3.1

Estimators of the volume fraction. Estimator (3.1) keeps its form for higher dimensions. Namely,

$$\hat{p}_W = \frac{\operatorname{mes}(\Xi \cap W)}{\operatorname{mes}(W)}$$

estimates the volume fraction of any stationary ergodic random closed set Ξ. Its unbiasedness follows from Robbins' theorem, since

$$\mathbf{E}\hat{p}_W = \frac{1}{\operatorname{mes}(W)} \int\limits_W \mathbf{P}\,\{x \in W\}\,dx = p\,.$$

The variance of $\hat{p}_W$ can be computed in the same way using second-order Robbins' theorem (2.10). Indeed,

$$\operatorname{Var}\hat{p}_W^2 = \frac{1}{\operatorname{mes}(W)^2} \int\limits_W \int\limits_W (C(x-y) - p^2)dxdy\,. \tag{3.5}$$

Girling's lemma. In the study of asymptotic variances the following lemma is often useful.

Lemma 3.1 (see [Gir82]) *If the integrable function $f : \mathbf{R}^d \mapsto \mathbf{R}$ depends on the difference of its arguments only, i.e. $f(x,y) = g(x-y)$ for a certain integrable over $\mathbf{R}^d$ function g, then*

$$\lim_{W \uparrow \mathbf{R}^d} \frac{1}{\operatorname{mes}(W)} \int\limits_W \int\limits_W f(x,y)dxdy = \int\limits_{\mathbf{R}^d} g(v)dv\,. \tag{3.6}$$

As we have already seen, such integrals appear in formula (3.5) for the second moment of the estimator $\hat{p}_W$.

Existence of the variance. For general d, (3.2) is valid if $\mathbf{E}\operatorname{mes}(\Xi_0)^2 < \infty$. To prove this, we use (2.18) to derive the following chain of inequalities:

$$
\begin{aligned}
\sigma^2 &= (1-p)^2 \int\limits_{\mathbf{R}^d} (\exp\{\lambda\mathbf{E}\operatorname{mes}(\Xi_0 \cap (\Xi_0 - v))\} - 1)dv \\
&\leq (1-p)^2 \exp\{\lambda\mathbf{E}\operatorname{mes}(\Xi_0)\} \int\limits_{\mathbf{R}^d} (1 - \exp\{-\lambda\mathbf{E}\operatorname{mes}(\Xi_0 \cap (\Xi_0 - v))\})dv
\end{aligned}
$$

$$\leq \quad \lambda(1-p) \int_{\mathbf{R}^d} \mathbf{E}\mathrm{mes}(\Xi_0 \cap (\Xi_0 - v))dv$$

$$= \quad \lambda(1-p)\mathbf{E} \left[\int_{\mathbf{R}^d} \int_{\mathbf{R}^d} 1_{\Xi_0}(x)1_{\Xi_0}(x+v)dxdv \right]$$

$$= \quad \lambda(1-p)\mathbf{E}\mathrm{mes}(\Xi_0)^2 \,.$$

Set-covariance function. The function

$$\gamma_F(x) = \mathrm{mes}(F \cap (F-x))$$

is said to be the set-covariance function of set F. A difficult problem of the unique determination of a set by its covariance function was discussed in [CJ94, LR90, Mat86, Nag93, Sch93a]. Note that the set-covariance function has appeared already in (2.18) and (3.3).

Meanwhile, sometimes characterisation problems can be solved by considering the Boolean model with a suitable grain and then using the uniqueness results for the Boolean model, see [LR90]. For example, assume that

$$\mathbf{E}\mathrm{mes}(X \oplus K) = \mathbf{E}\mathrm{mes}(Y \oplus K) \tag{3.7}$$

for all compact sets K and two random compact sets X and Y. Then X and Y have the same distributions. Indeed, consider two Boolean models with the same intensities and the typical grains given by X and Y. Then (3.7) implies that these Boolean models have the same distribution, whence the distributions of their grains also coincide, see p. 23.

Asymptotic normality. In fact, the asymptotic normality of the estimator $\hat{p}_W$ can be derived from the mixing (or weak dependence) property of any stationary random set Ξ. Such an approach was used in [Mas82] to formulate the corresponding central limit theorem.

In the case of the Boolean model it is also possible to reduce the problem to the study of m-dependent random fields by truncating the grain. For this, consider a new Boolean model $\Xi^{(r)}$ with the same germs but with 'truncated' grain $\Xi_0^{(r)} = \Xi_0 \cap B_r(o)$ for some $r > 0$. Then the indicator function of $\Xi^{(r)}$ is an m-dependent random field, that is, it has independent values if the distance between arguments is greater than $2r$. Thus, a limit theorem for m-dependent random fields (see, e.g., [Hei88a, Hei93]) yields the asymptotic normality of the estimator $\hat{p}_W^{(r)}$ of the volume fraction, $p^{(r)}$, of the set $\Xi^{(r)}$. To this end, it is sufficient to show that the rest of the Boolean model (the set $\Xi \setminus \Xi^{(r)}$) plays no role, i.e. for the given family of windows $W_s, s \geq 1$,

$$\lim_{r \to \infty} \sup_{s \geq 1} \mathrm{mes}(W_s)\mathrm{Var}\,(\hat{p}_{W_s} - \hat{p}_{W_s}^{(r)}) = 0 \,.$$

This was proved in [Hei93] even for the more general case of contact distribution functions (see Section 4.3) provided $\mathbf{E}\mathrm{mes}(\Xi_0^r)^2 < \infty$ for some $r > 0$.

Stereological notations and identities. Stereological notations are used to denote specific values of certain parameters of interest. We have already seen that A_A denotes the mean area (upper A) of the set in question per unit area (subscript A) in the plane. Furthermore, V, L, J, P, N, S (also as indices) denote respectively the volume, the curve length in the plane, the curve length in the space, the number of points, the number of particles, and the surface area.

The identity

$$V_V = A_A = L_L = P_P \tag{3.8}$$

gives different possibilities to estimate the volume fraction. In fact, (3.8) easily follows from stationarity and is valid for each stationary random closed set Ξ, see [SKM95, pp. 342, 343]. There are also many other stereological identities, see [Wei80] and [SKM95, Chapter 11]. For example,

$$
\begin{aligned}
S_V &= 4L_A/\pi = 2P_L\,, \\
J_V &= 2P_A\,.
\end{aligned}
$$

3.2   Specific Convexity Number

Let us associate with the grain Ξ_0 its tangent point $n_u(\Xi_0)$ in the direction u (u is a vector from the unit circle $\mathbf{S}^1$ in the plane or from the unit sphere $\mathbf{S}^{d-1}$ in the space of dimension d), see Figure 3.1. If there are several tangent points, then we choose one according to a fixed rule, say with the largest ordinate. If the vector u is directed upwards, then we will get the *lower positive tangent point* of Ξ_0 as discussed in [MS94a].

Let us consider the set of such tangent points determined for all shifted grains. Some of these points are covered by other grains. The *exposed tangent points* form the point process denoted by $N^+(u)$. Sometimes these tangent points are said to be *positive* in contrast to those (negative) tangent points produced by the complement of Ξ. Figure 3.2 shows exposed tangent points marked by crosses.

We can redefine our original Boolean model using all tangent points as germs and the shifted grain

$$\Xi_0^u = \Xi_0 - n_u(\Xi_0) \tag{3.9}$$

as the new 'typical' grain [MS94a, Mol95]. This new shifted grain has its tangent point at the origin. Then all tangent points form a Poisson point process of germs with the same intensity λ, while the *exposed* tangent points are now obtained as a result of a *dependent thinning* procedure applied to the Poisson point process of germs.

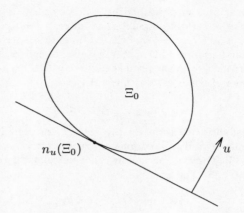

Figure 3.1   A tangent point of Ξ_0.

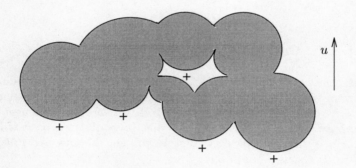

Figure 3.2   Exposed lower positive tangent points for u directed upwards.

It is possible to show [SKM95, p. 78] that the intensity of the process $N^+(u)$ does not depend on u and is equal to

$$N_A^+ = \lambda(1 - p).$$
(3.10)

This value is said to be the *specific convexity number* of Ξ. (In general, the convexity number is the average number of exposed tangent points for all directions.) The estimation of N_A^+ can be performed in the following way. First, choose an arbitrary direction u and count the number $\mathsf{N}(N^+(u) \cap W)$ of tangent points in the direction u inside the window W. Then

$$\hat{N}_{A,W}^+ = \frac{\mathsf{N}(N^+(u) \cap W)}{\mathsf{A}(W)}$$
(3.11)

is a strong consistent asymptotically normal estimator of N_A^+. In other words, $\hat{N}_{A,W}^+$ converges to N_A^+ with probability one and their normalised difference, $A(W)^{1/2}(\hat{N}_{A,W}^+ - N_A^+)$, converges in distribution to the centred Gaussian random variable with the variance

$$
\begin{aligned}
V_N(u) &= \lim_{W \uparrow \mathbf{R}^2} A(W) \operatorname{Var} \hat{N}_{A,W}^+ \\
&= \lambda(1-p) + \lambda^2(1-p)^2 \int_{\mathbf{R}^2} (g_u(v) - 1)\,dv,
\end{aligned}
\tag{3.12}
$$

where

$$
g_u(v) = q(v)(\psi_u(v) + \psi_u(-v) - 1)
\tag{3.13}
$$

is the pair-correlation function (see Section 4.4, [SKM95, p. 129] and [SS94, p. 249]) of the point process $N^+(u)$, and the function

$$
\psi_u(v) = \mathbf{P}\{v \notin \Xi_0^u\} = 1 - \mathbf{P}\{v \in \Xi_0^u\}
\tag{3.14}
$$

is given by the covering probabilities of the shifted grain (3.9). The function q is related to the covariance of Ξ and defined by (2.18).

If the Boolean model is isotropic, then $V_N(u)$ does not depend on u. If, in addition, the covariance is exponential, then

$$
V_N = \lambda(1-p) + 2\pi^3 p^3 (1-p)^3 \lambda^2 L_A^{-2},
$$

i.e. V_N can be obtained through the numerical parameters of Ξ.

It should be noted that the estimator (3.11) is not always the best (efficient) estimator of the intensity of the point process $N^+(u)$. A further problem in implementation of the tangent points counting on computers is caused by missing some tangent points as a result of digitising, see [Cre91, p. 769] and [MS94a, Sch92]. This leads to underestimation effects which are noticeable for Boolean models with high intensities and 'small' grains.

Notes to Section 3.2

Intensity of the exposed tangent points. The idea of the proof of (3.10) can be explained as follows, see [SKM95, p. 78]. First, it is possible to use all (both exposed and covered) tangent points as germs and the shifted grain Ξ_0^u as the typical grain to build the Boolean model with the same distribution as the original Boolean model Ξ. Then the germ process is a stationary Poisson point process of intensity λ, since all tangent points are obtained by independent shifts of the points belonging to the original Poisson point process of germs (here the independence of grains is important). Each of the new germ points x_i is exposed and, therefore, belongs to $N^+(u)$, if it is not covered by all other grains. By Slivnyak's theorem [DVJ88, p. 459], the union of all grains $\Xi_j + x_j$ with $j \neq i$ has the same distribution as Ξ. Therefore, x_i is not covered by all *other* grains with probability $(1 - p)$. Now the Campbell theorem implies that the intensity of the point process of exposed tangent points is $\lambda(1 - p)$.

If the germ process is *not* Poisson, then it is possible to derive the intensity of exposed tangent points as

$$N_A^+(u) = \lambda G_0^! \left[\mathbf{P} \left\{ (\cdot) \notin \check{\Xi}_0^u \right\} \right],$$

see [HM95], where $G_0^!$ is the probability generating functional of the point process of germs taken with respect to its reduced Palm distribution, see [DVJ88, p. 455].

Convexity number. The convexity number $\bar{V}_0(F)$ of a set F from the convex ring is defined by (2.27). On the other hand, $\bar{V}_0(F)$ can be computed as the average number of tangent points of F for all directions, see [Ser82, p. 140],

$$\bar{V}_0(F) = \frac{1}{\omega_{d-1}} \int_{\mathbf{S}^{d-1}} N^+(u) du,$$

where ω_{d-1} is the surface area of $\mathbf{S}^{d-1}$. Then (3.10) yields also

$$\frac{\mathbf{E}\bar{V}_0(\Xi \cap W)}{\text{mes}(W)} = \lambda(1 - p).$$

Formal definition of the tangent point. For a convex compact set K and a direction $u \in \mathbf{S}^{d-1}$ define the *support set*

$$\partial_u K = \{x \in \partial K : \langle u, x \rangle = -h(K, u)\},$$

where ∂K is the boundary of K and $h(K, u)$ is the support function of K, see (2.15) and Figure 2.5. The set $\partial_u K$ may contain either a single point or some subset of the boundary ∂K. We define $n_u(K)$ to be the lexicographical minimum of $\partial_u K$. If $\partial_u K$ is a singleton, then $\partial_u K = \{n_u(K)\}$. The point $n_u(K)$ is said to be the tangent point of K in the direction u. Note that (3.10) remains valid for an arbitrary (the same for all grains and determined only by their support sets) choice of a point $n_u(K) \in \partial_u K$.

Positive boundary. The positive boundary of a set F is defined to be the set of all points $z \in \partial F$, such that there exists a point y with z being its metric projection on F, see p. 27. The positive boundary is denoted by $\partial_+ F$. If F belongs to the convex ring and $F = \cup K_i$ for convex sets $K_1, \ldots, K_n$, then

$$\partial_+ K = \bigcup_{i=1}^{n} (\partial K_i \setminus \cup_{j \neq i} K_j)$$

is the union of all exposed parts of boundaries of $K_1, \ldots, K_n$. Then $\partial_+ F$ is the union of all possible exposed tangent points of the sets $K_1, \ldots, K_n$ and $\partial F \setminus \partial_+ F$ is comprised of parts with dimensions less than or equal to $(d - 2)$. Note that the positive extensions $\bar{V}_j(F)$, $0 \leq j \leq d-1$, of intrinsic volumes (see p. 27) depend on the positive boundary of F only.

The subset of the boundary of F comprising the support sets $\partial_u F$ with $u \in \Gamma \subset \mathbf{S}^{d-1}$ is given by

$$\partial_\Gamma F = \bigcup_{u \in \Gamma} \bigcup_{i=1}^{n} (\partial_u K_i \setminus \cup_{j \neq i} K_j).$$

Clearly, $\partial_\Gamma F \subset \partial_+ F$. If F is convex, then $\partial_\Gamma F$ is called the reverse spherical image of F, see [Sch93b, p. 77].

Product densities of the point process. Let us consider a general point process N in $\mathbf{R}^d$. Its nth-order *moment measure* is defined by

$$\begin{aligned}
\mu^{(n)}(W_1 \times \cdots \times W_n) &= \mathbf{E}[N(W_1) \cdots N(W_n)] \\
&= \mathbf{E} \sum_{x_1, \ldots, x_n \in N} \mathbf{1}_{x_1 \in W_1} \cdots \mathbf{1}_{x_n \in W_n}
\end{aligned}$$

for all Borel sets $W_1, \ldots, W_n$. Note that N designates at the same time the realisation of the point process and the corresponding counting measure.

If the sum is taken over the family of disjoint points, then we get the *factorial moment measure* of N,

$$\alpha^{(n)}(W_1 \times \cdots \times W_n) = \mathbf{E} \sum_{\substack{x_1, \ldots, x_n \in N \\ x_i \neq x_j, \, 1 \leq i, j \leq n}} \mathbf{1}_{x_1 \in W_1} \cdots \mathbf{1}_{x_n \in W_n}. \tag{3.15}$$

The density $\rho^{(n)}$ of $\alpha^{(n)}$ is said to be the *product density* of N. The second-order factorial moment measure is especially important. In the stationary case its density, $\rho^{(2)}$, depends on the difference $x - y$ only. The normalised density

$$g(v) = \lambda^{-2} \rho^{(2)}(o, v) \tag{3.16}$$

is said to be the *pair-correlation function* of N. Here λ is the intensity of the point process N.

The variance of the number of points inside a window W is then easy to calculate by

$$\operatorname{Var} N(W) = \lambda \operatorname{mes}(W) + \lambda^2 \int_W \int_W (g(x - y) - 1) dx dy,$$

see, e.g., [SS94, pp. 247–248].

Point process of exposed tangent points. Denote by

$$\tilde{x}_i = n_u(\Xi_i + x_i) = n_u(\Xi_i) + x_i$$

the tangent point of the ith shifted grain. The tangent points process $N^+(u)$ is defined by

$$N^+(u) = \left\{ \tilde{x}_i \colon i \geq 1, \, \tilde{x}_i \notin \bigcup_{j \neq i} (\Xi_j^u + \tilde{x}_j) \right\}.$$

If u is directed upwards, then $N^+(u)$ is exactly the process of lower positive tangent points considered in [MS94a]. We denote by $N^+(u, W) = \mathsf{N}(N^+(u) \cap W)$ the number of points of $N^+(u)$ inside the window W.

The point process $N^+(u)$ is stationary, but *anisotropic* (in contrast to the original Poisson germ process). For example, in the point process of lower positive tangent points the presence of an exposed tangent point at x causes the absence of other points above and below x, but not to the right or to the left. Although the intensity of $N^+(u)$ is the same for all u, its *distribution* does depend on u. In general, the distributions for different u cannot be made equal by means of rotations.

The following result gives the pair-correlation function and the higher-order product densities of the point process $N^+(u)$.

Theorem 3.2 (see [MS94a]) *The pair-correlation function of $N^+(u)$ exists and is given by*

$$g_u(v) = q(v)(\psi_u(v) + \psi_u(-v) - 1) = q(v)\mathbf{P}\left\{\{v, -v\} \cap \Xi_0^u = \emptyset\right\}. \qquad (3.17)$$

Furthermore, the nth-order product density of the point process $N^+(u)$ equals

$$\rho_u^{(n)}(x_1, \ldots, x_n) = \lambda^n Q_\Xi(\{x_1, \ldots, x_n\}) \qquad (3.18)$$

$$\times \prod_{j=1}^{n} \psi_u(x_1 - x_j, \ldots, x_{j-1} - x_j, x_{j+1} - x_j, \ldots, x_n - x_j),$$

where $Q_\Xi(\{x_1, \ldots, x_n\}) = 1 - T_\Xi(\{x_1, \ldots, x_n\})$ and

$$\psi_u(x_1, \ldots, x_n) = \mathbf{P}\left\{\{x_1, \ldots, x_n\} \cap \Xi_0^u = \emptyset\right\}. \qquad (3.19)$$

Tangent points in different directions. The processes $N^+(u_i)$, $1 \leq i \leq m$, can be drawn for different directions $u_1, \ldots, u_m$. Let $N^+(u_1, \ldots, u_m)$ be the *marked point process* of tangent points with the mark space $\{u_1, \ldots, u_m\} \subset \mathbf{S}^{d-1}$. Its points (with the marks stripped) are determined by the union of $N^+(u_i)$ for $i = 1, \ldots, m$, and each point x has the mark u_j if $x \in N^+(u_j)$. The second-order characteristics of this marked point process were computed in [Mol95].

Theorem 3.3 (see [Mol95]) *For each Borel set W_1, W_2 and $u_1, u_2 \in \mathbf{S}^{d-1}$,*

$$\mathbf{E} N^+(u_1, W_1) N^+(u_2, W_2)$$

$$= \lambda^2 (1-p)^2 \int\limits_{W_1} \int\limits_{W_2} q(y-x)\psi_{u_1}(y-x)\psi_{u_2}(x-y)dxdy$$

$$+ \lambda(1-p)^2 \int\limits_{\mathbf{R}^d} q(v)\mathrm{mes}(W_1 \cap (W_2 - v))P_{u_1, u_2}(dv), \qquad (3.20)$$

where $\psi_u(v) = \mathbf{P}\{v \notin \Xi_0^u\}$ and $P_{u_1, u_2}(.)$ is the distribution of the random point $\xi_{u_1, u_2} = n_{u_2}(\Xi_0^{u_1})$.

Theorem 3.3 immediately yields the pair-correlation function of the marked point process of tangent points.

Theorem 3.4 (see [Mol95]) *Suppose that the distribution P_{u_i,u_j} admits the density p_{u_i,u_j}. Then the pair-correlation function of the point process $N^+(u_1, \ldots, u_m)$ between points having the marks u_i and u_j is given by*

$$g_{u_i,u_j}(v) = q(v)\psi_{u_i}(v)\psi_{u_j}(-v) + \lambda^{-1}p_{u_i,u_j}(v)q(v)1_{i\neq j} . \tag{3.21}$$

3.3 Specific Boundary Length

The specific boundary length, L_A, of Ξ is its expected boundary length per unit area. Therefore, the measurement of the specific boundary length is based on the measurement of the boundary length of planar sets. Note that such a measurement is not very easy to implement on the digitised computer screen. The corresponding technical problems are discussed in [Ser82, pp. 220–223].

Since, in fact, all measurements are performed within a bounded window, we can measure the boundary length $U(\Xi \cap W)$ of the set $\Xi \cap W$. This boundary length, is the sum of two parts:

- The boundary length of Ξ inside the *interior*, $\operatorname{Int} W$, of W: $U(\Xi \cap \operatorname{Int} W)$.
- The length of the boundary, ∂W, of W covered by Ξ: $U(\Xi \cap \partial W)$.

The value

$$U(\Xi \cap W) = U(\Xi \cap \operatorname{Int} W) + U(\Xi \cap \partial W)$$

is said to be the *total boundary length* of $\Xi \cap W$.

Therefore, the specific boundary length, L_A, of Ξ can be estimated either by

$$\hat{L}^i_{A,W} = \frac{U(\Xi \cap \operatorname{Int} W)}{A(W)} \tag{3.22}$$

or by

$$\hat{L}^b_{A,W} = \frac{U(\Xi \cap W)}{A(W)} . \tag{3.23}$$

Both estimators (3.22) and (3.23) are strong consistent, but biased. In practice it is impossible to measure lengths inside the *interior* of W. The following unbiased estimator, [WW84]

$$\hat{L}_{A,W} = \frac{U(\Xi \cap W) - U(\Xi \cap \partial^+ W)}{A(W)} , \tag{3.24}$$

is often easier to compute. Here $\partial^+ W$ is the *upper-right* boundary of the rectangular window W and $U(\Xi \cap \partial^+ W)$ is the length of the corresponding intersection.

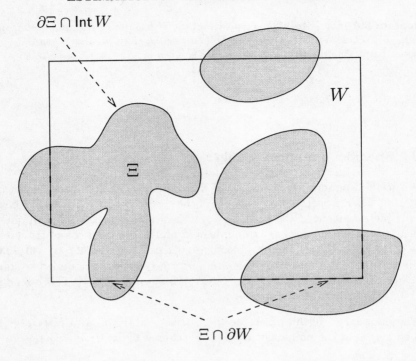

Figure 3.3 Two parts of the boundary of $\Xi \cap W$.

The boundary $\partial\Xi$ of Ξ is a particular case of the so-called *fibre process* (or surface process in higher dimensions). In general, a fibre process is a union of one-dimensional curves (fibres) [SKM95, Chapter 9]. The specific boundary length is called the intensity of the fibre process $\partial\Xi$. Estimation methods for its intensity and other parameters (the rose of intersections, the second moment function, etc.) are discussed in [SKM95, pp. 292–296] and [MS94b], see also [Wei80, Chapter 3] and Section 3.5. Note that the characteristics of the fibre process $\partial\Xi$ belong to aggregate parameters, since this fibre process is completely observable.

The total length of the boundary can be measured directly by approximations by straight lines (if an image analyser or relevant software is available). Otherwise it is possible to use indirect methods, e.g.,

$$\hat{L}_{A,W} = \frac{\pi \int\limits_{0}^{2\pi} \mathsf{N}(\Phi \cap \omega\Phi_0 \cap W)d\omega}{2 \int\limits_{0}^{2\pi} \mathsf{U}(\omega\Phi_0 \cap W)d\omega} \tag{3.25}$$

estimates L_A, where $\omega \Phi_0$ is the rotation by the angle ω of the test systems Φ_0 of parallel segments. The integral in the numerator is the integrated number of intersection points between Φ and the rotated test system Φ_0 within W. The integral in the denominator is the integrated length of Φ_0 visible from the window W. Equation (3.25) follows from the formula [SKM95, p. 292] for the intensity of the intersection of two fibre processes.

In the isotropic case (3.25) can be replaced by

$$\hat{L}_{A,W} = \frac{\pi N(\Phi \cap \Phi_0)}{2U(\Phi_0)} \qquad (3.26)$$

for $\Phi_0 \subset W$. If Φ_0 is a test system of circles, then (3.26) is applicable in both isotropic and anisotropic cases.

Another important fibre process related to the Boolean model is defined as the union of the boundaries of all shifted grains as though the grains themselves were transparent:

$$\partial_f \Xi = \bigcup_{i:\, x_i \in \Pi_\lambda} (x_i + \partial \Xi_i). \qquad (3.27)$$

The intensity of this fibre process is equal to $\lambda EU(\Xi_0)$. However, the fibre process $\partial_f \Xi$ is not observable because of occlusions.

Notes to Section 3.3

Second-order characteristics of fibre processes. Let Φ be a stationary fibre process on the plane, see [SKM95, p. 282]. If $\Phi(W)$ is the total length of Φ inside W, then $\mathbf{E}\Phi(W) = L_A \mathrm{mes}(W)$, where L_A is said to be the intensity of W. Furthermore, the stationarity property yields the decomposition

$$\mathbf{E}[\Phi(W_1)\Phi(W_2)] = L_A^2 \int_{\mathbf{R}^2} \int_{\mathbf{R}^2} \mathbf{1}_{W_1}(x)\mathbf{1}_{W_2}(v+x)dx\mathcal{K}(dv). \qquad (3.28)$$

The measure $\mathcal{K}(\cdot)$ is said to be the reduced second moment measure of Φ. Its density (if it exists) is denoted by g_Φ and said to be the *pair-correlation function* of Φ, see also [SKM95, p. 284]. Then (3.28) yields a simple expression for the variance of $\Phi(W)$,

$$\mathrm{Var}\, \Phi(W) = L_A^2 \left[\int_W \int_W (g_\Phi(x-y) - 1)dxdy \right].$$

By Lemma 3.1, the asymptotic variance is equal to

$$\lim_{W \uparrow \mathbf{R}^2} A(W)^{-1}\mathrm{Var}\, \Phi(W) = L_A^2 \int_{\mathbf{R}^2} (g_\Phi(v) - 1)dv \qquad (3.29)$$

if the corresponding integral exists, see also [Sto83].

Asymptotic variances of estimators. Note that the truncation argument as given on p. 32 is not applicable when proving limit theorems for the specific surface area or boundary length. In general, the limit theorem for β-mixing random fields [Hei94b] or more sophisticated approximations by m-dependent random fields [HM95] must be applied. For simplicity, we suppose that the typical grain is contained almost surely in a certain fixed ball. Then the corresponding fibre process $\partial\Xi$ is m-dependent, whence (3.22) yields an asymptotically normal estimator with the limiting variance given by

$$V_L = L_A^2 \int_{\mathbf{R}^2} \Big(g_{\partial\Xi}(v) - 1 \Big) dv \tag{3.30}$$

for each $u \in \mathbf{S}^1$ [MS94a]. In fact, (3.30) is a particular case of (3.29) applied to the fibre process $\partial\Xi$. The pair-correlation function $g_{\partial\Xi}$ can be found through the pair-correlation function $g_f(\cdot)$ of the fibre process $\partial_f\Xi$ as

$$g_{\partial\Xi}(v) = q(v)(\psi_u(v) + \psi_u(-v) - 1)g_f(v) \, .$$

However, formula (3.30) for the variance is of little practical use, since V_L is expressed through unknown individual parameters of the Boolean model. Estimation problems of the pair-correlation function were considered in [Sto85].

Specific surface areas. It is possible either to define spatial densities for parts of the boundary that satisfy some properties (e.g., consist of points with given outer normal vectors). To be general enough, introduce the set

$$B_r(F; \Gamma) = \bigcup_{u \in \Gamma} \bigcup_{x \in \partial_u F} \{x + tu : \ 0 \le t \le r\} \, ,$$

where F belongs to the convex ring and Γ is a Borel subset of the unit sphere $\mathbf{S}^{d-1}$. Similarly to the Steiner formula (2.26),

$$\operatorname{mes}(B_r(F; \Gamma)) = \sum_{j=0}^{d-1} r^{d-j} b_{d-j} S_j(F; \Gamma) \, . \tag{3.31}$$

Then $S_{d-1}(F; \Gamma)$, $\Gamma \subset \mathbf{S}^{d-1}$, is called the surface area measure. In the smooth case $S_{d-1}(F; \Gamma)$ can be interpreted as the $(d-1)$-dimensional Hausdorff measure of the set of all points of ∂F with outer normal vectors belonging to Γ. Clearly, $S_{d-1}(F, \mathbf{S}^{d-1})$ is the surface area of F. Then

$$\frac{\mathbf{E}S_{d-1}(\Xi \cap W; \Gamma)}{\operatorname{mes}(W)} = \lambda(1-p)\mathbf{E}S_{d-1}(\Xi_0; \Gamma) + \frac{pS_{d-1}(W; \Gamma)}{\operatorname{mes}(W)} \, ,$$

see [Sch92, Wei95], whence the density of $S_{d-1}(\cdot; \Gamma)$ is given by

$$D_{S_{d-1}}(\Xi; \Gamma) = \lambda(1-p)\mathbf{E}S_{d-1}(\Xi_0; \Gamma) \, .$$

A central limit theorem for estimators of surface area densities was proved in [Mol95]. For simplicity, assume that Ξ_0 a.s. contains no flat pieces on its boundary.

Theorem 3.5 (see [Mol95]) *If $\Xi_0 \subset K_0$ a.s. for a fixed compact set K_0 and $W \uparrow \mathbf{R}^d$ is an expanding sequence of open windows, then, for all Borel $\Gamma \subset \mathbf{S}^{d-1}$,*

$$
\text{mes}(W)^{1/2} \left(\frac{S_{d-1}(\Xi \cap W; \Gamma)}{\text{mes}(W)} - D_{S_{d-1}}(\Xi; \Gamma) \right)
$$

converges in distribution to a centred Gaussian random variable with the variance

$$
V_\Gamma = \int_\Gamma \int_\Gamma \left[\lambda^2 (1-p)^2 \int_{\mathbf{R}^d} \left[q(v) \Psi_{u_1}(v; du_1) \Psi_{u_2}(-v; du_2) \right. \right.
$$
$$
\left. - \mathbf{E}S_{d-1}(\Xi_0; du_1) \mathbf{E}S_{d-1}(\Xi_0; du_2) \right] dv
$$
$$
\left. + \lambda (1-p)^2 \mathbf{E} \left[q(\xi_{u_1, u_2}) S_{d-1}(\Xi_0; du_1) S_{d-1}(\Xi_0; du_2) \right] \right],
$$

where

$$
\xi_{u_1, u_2} = n_{u_1}(\Xi_0) - n_{u_2}(\Xi_0)
$$

is the difference between two tangent points (a chord of Ξ_0), and

$$
\Psi_u(v; du) = \mathbf{E} \left[\mathbf{1}_{v \notin \Xi_0^u} S_{d-1}(\Xi_0; du) \right].
$$

A more general case of random measures associated with the Boolean model is considered in [HM95]. A broad generalisation of fibre (and surface) processes was introduced and studied in [Zäh82, Zäh86].

3.4 Specific Connectivity Number

The *connectivity number* χ is also known under the name of Euler–Poincaré characteristic. Roughly speaking, $\chi(F)$ is equal to the difference between the number of clumps and the number of holes formed by planar compact set F. Formally, for a convex compact set F the value $\chi(F)$ is 1 if F is non-void and 0 otherwise. If $F = F_1 \cup F_2$ for two convex sets F_1 and F_2, then $\chi(F)$ is additively extended by

$$
\chi(F) = \chi(F_1) + \chi(F_2) - \chi(F_1 \cap F_2).
$$

The values on $F = F_1 \cup \cdots \cup F_n$ are defined similarly.

Clearly, $\chi(F) = n$ if F can be represented as a union of n disjoint convex compact sets. In this case the number of tangent points (the convexity number) is also equal to n. However, in general $\chi(F)$ and $\mathsf{N}^+(F)$ are not equal. For instance, the number of tangent points is always positive (for a non-empty set), while the connectivity number can be negative as well.

For the determination of the connectivity number for planar sets the technique of [Kel85] can be applied. It is based on the formula

$$\chi(\Xi \cap W) = \mathsf{N}(N^+(u) \cap W) - \mathsf{N}(N^-(-u) \cap W), \qquad (3.32)$$

which states that the connectivity number is equal to the difference between the number of tangent points of Ξ (positive tangent points) for any direction u and the number of tangent points of the complement of Ξ in the direction $-u$ (negative tangent points), see also [SKM95, pp. 240–242]. Figure 3.4 illustrates (3.32) with the points of the first kind marked by '+' and of the second kind by '−'.

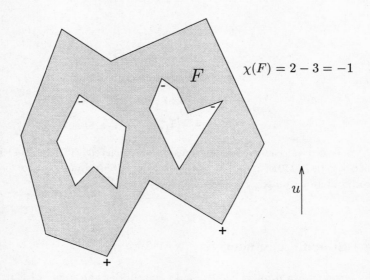

Figure 3.4   The evaluation of the connectivity number.

Theoretical formulae for the *specific* connectivity number χ_A (the expected connectivity number per unit area) of the Boolean model were obtained in [Kel83, Kel84, Mil76, WW84, Wei88]. In the simplest case χ_A is given by (2.14).

An estimator, $\hat{\chi}_{A,W}(u)$, of the specific connectivity number can be obtained by dividing $\chi(\Xi \cap W)$ from (3.32) by $\mathsf{A}(W)$. To get an unbiased estimator of the specific connectivity number the same boundary correction as in (3.24) should be applied: namely, an unbiased estimator of the specific connectivity number is given by

$$\hat{\chi}_{A,W} = \frac{\chi(\Xi \cap W) - \chi(\Xi \cap \partial^+ W)}{\mathsf{A}(W)}, \qquad (3.33)$$

where $\partial^+ W$ is the upper-right boundary of W, see Figure 3.5. The estimator (3.33) is strong consistent. Unfortunately, its asymptotic variance is unknown, although some preliminary results can be found in [HM95].

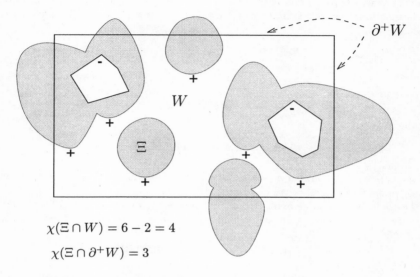

$$\chi(\Xi \cap W) = 6 - 2 = 4$$
$$\chi(\Xi \cap \partial^+ W) = 3$$

Figure 3.5 Unbiased estimation of the connectivity number.

These estimators can be averaged for different directions to achieve lower variances. In practice, the estimator of χ_A based on (3.32) behaves reasonably well. The usual underestimation effects in the tangent points counting (see Section 3.2) compensate each other after the subtraction in (3.32).

The digitised connectivity number (discrete analogue of χ for subsets of the grid) depends on the structure of the grid (or graph) used to define the connectivity. For example, in the quadratic grid

$$\chi^{iv}(F) = \mathsf{N} \left\{ \begin{matrix} 1 & 0 \\ 0 & 0 \end{matrix} \right\} + \mathsf{N} \left\{ \begin{matrix} 1 & 0 \\ 0 & 1 \end{matrix} \right\} - \mathsf{N} \left\{ \begin{matrix} 1 & 1 \\ 1 & 0 \end{matrix} \right\}, \qquad (3.34)$$

where $\mathsf{N}\{\cdot\}$ denotes the number of the corresponding pixel configurations in the digitised set F, see [Ser82, p. 201]. If the grid (graph) is 8-connected, then

$$\chi^{viii}(F) = \mathsf{N} \left\{ \begin{matrix} 1 & 0 \\ 0 & 0 \end{matrix} \right\} - \mathsf{N} \left\{ \begin{matrix} * & 1 \\ 1 & 0 \end{matrix} \right\}, \qquad (3.35)$$

where '*' represents either 0 or 1. A general approach to digitising of some geometric functionals is suggested in [Hei92a, Hei94a].

Notes to Section 3.4

Intersection with a semi-open cube. Weil and Wieacker [WW84] established the existence of spatial densities of locally bounded measurable additive translation-invariant functionals on the convex ring. For the case of the Boolean model their result gives

$$D_\phi(\Xi) = \lim_{W \uparrow \mathbf{R}^d} \frac{\mathbf{E}\phi(\Xi \cap W)}{\mathrm{mes}(W)} = \mathbf{E}[\phi(\Xi \cap W_0) - \phi(\Xi \cap \partial^+ W_0)], \qquad (3.36)$$

where

$$W_0 = \{x = (x_1, \ldots, x_d) : \ 0 \leq x_i \leq 1, \ 1 \leq i \leq d\}$$

is the unit cube and

$$\partial^+ W_0 = \{x \in W_0 : \ \max_{1 \leq i \leq d}(x_i) = 1\}$$

is its upper-right boundary. Now (3.33) follows from (3.36) if ϕ is taken to be the Euler–Poincaré characteristic.

Formula (3.36) can be explained in the following simple way. Let $W^0 = W \setminus \partial^+ W$ denote the *semi-open* unit cube. Then take $W = [-n, n)^d$ and note that, by additivity,

$$\phi(\Xi \cap [-n, n)^d) = \sum_{x \in [-n,n)^d \cap \mathbb{Z}^d} \phi(\Xi \cap (W^0 + x)),$$

where the sum is taken with respect to all x with integer-valued coordinates. Then (3.36) follows from the spatial ergodic theorem in [NZ79].

Densities of geometric functionals. Let ϕ be an additive translation-invariant functional on the convex ring that is locally bounded and homogeneous of degree j, $0 \leq j \leq d$. The homogeneity means that $\phi(cK) = c^j \phi(K)$ for $c > 0$ and all K from the convex ring. For example, the connectivity number is homogeneous of degree 0, while the surface area is homogeneous of degree $(d-1)$.

Furthermore, suppose that the restriction of ϕ on the family of convex sets is continuous with respect to the Hausdorff metric. It was proved in [WW84] that, for a convex set W,

$$\mathbf{E}\phi(\Xi \cap W) = \phi(\{o\})\mathbf{P}\{o \in \Xi\} + \sum_{i=j+1}^{d-1} \phi_i(W) + D_\phi(\Xi)\mathrm{mes}(W),$$

where ϕ_i, $j + 1 \leq i \leq d$, are continuous translation-invariant functionals on the family of convex sets, which depend on Ξ in such a way that

$$\phi_i(K) = \phi_i(\Xi, K) \quad \text{with} \quad \phi_i(c\Xi, K) = c^{d+j-i}\phi_i(\Xi, K)$$

for all $c > 0$. Under additional assumptions this result is valid also in a more general setting of stationary random sets from the extended convex ring, see [WW84].

The Boolean model assumption yields a more explicit representation for $\mathbf{E}\phi(\Xi \cap W)$, see [WW84]. Let Θ be a measure on $\mathcal{K}$ defined by $\Theta(\mathcal{K}_W) = \lambda \mathbf{E}\mathrm{mes}(\Xi_0 \oplus \check{W})$

for $\mathcal{K}_W = \{K \in \mathcal{K} : K \cap W \neq \emptyset\}$. Then the number of shifted grains $(x_i + \Xi_i)$ that hit W has the Poisson distribution with parameter $\Theta(\mathcal{K}_W)$ and each such grain hitting W has the distribution $\Theta(\cdot)/\Theta(\mathcal{K}_W)$. Thus,

$$\mathbf{E}\phi(\Xi \cap W)$$

$$= \exp\{-\Theta(\mathcal{K}_W)\} \sum_{n=1}^{\infty} \frac{1}{n!} \int_{\mathcal{K}} \cdots \int_{\mathcal{K}} \phi(W \cap (K_1 \cup \cdots \cup K_n))\Theta(dK_1)\cdots\Theta(dK_n).$$

Put

$$w(p) = \int_{\mathcal{K}} \cdots \int_{\mathcal{K}} \phi(W \cup K_1 \cup \cdots \cup K_p)\Theta(dK_1)\cdots\Theta(dK_p), \quad p \geq 1.$$

Then

$$\mathbf{E}\phi(\Xi \cap W) = \exp\{-\Theta(\mathcal{K}_W)\} \sum_{n=1}^{\infty} \frac{1}{n!} \sum_{p=1}^{n} (-1)^{p-1} \binom{n}{k} [\Theta(\mathcal{K}_W)]^{n-p} w(p).$$

It is possible to justify the change of order of the two sums, whence

$$\mathbf{E}\phi(\Xi \cap W) = \sum_{p=1}^{\infty} \frac{(-1)^{p-1}}{p!} \int_{\mathcal{K}} \cdots \int_{\mathcal{K}} \phi(W \cup K_1 \cup \cdots \cup K_p)\Theta(dK_1)\cdots\Theta(dK_p).$$

The intensity measure Θ can be decomposed as the product of the Lebesgue measure on $\mathbf{R}^d$ and probability measure $Q(\cdot)$ on $\mathcal{K}$, which determines the distribution of the typical grain. Thus,

$$\mathbf{E}\phi(\Xi \cap W) = \sum_{p=1}^{\infty} \frac{(-1)^{p-1}}{p!} \int_{\mathcal{K}} \cdots \int_{\mathcal{K}} \tag{3.37}$$

$$\int_{\mathbf{R}^d} \cdots \int_{\mathbf{R}^d} \phi(W \cup (K_1 + x_1) \cup \cdots \cup (K_p + x_p))dx_1 \cdots dx_p \Theta(dK_1)\cdots\Theta(dK_p).$$

Further formulae for $\phi = V_j$ (i.e. for densities of intrinsic volumes, see p. 27) can be obtained from the iterated kinematic integral formula in the isotropic case [WW84] or the iterated translative integral geometric formula in the general case [Wei88, Wei90]. For instance, the kinematic formula gives

$$\int_{SO_d} \int_{\mathbf{R}^d} V_j(K \cap g(L + x))dx d\mu(g) = \sum_{k=j}^{d} \alpha_{djk} V_k(K) V_{d+j-k}(L) \tag{3.38}$$

for two convex sets K and L [Sch93b, p. 253], [SW92, SW93]. Here SO_d is the group of rotations, μ is the Haar measure on SO_d such that $\mu(SO_d) = 1$, and

$$\alpha_{djk} = \frac{\binom{k}{j} b_k b_{d+j-k}}{\binom{d}{k-j} b_j b_d}$$

(remember that b_d is the volume of the unit ball in $\mathbf{R}^d$). The translative integral geometric formula deals with integration over $\mathbf{R}^d$ only.

To compute the right-hand side of (3.37) iterations of the integral-geometric formulae [Str70, Wei90] are needed. For instance, if the Boolean model is isotropic, then

$$
\begin{aligned}
\mathbf{E}V_j(\Xi \cap W) &= V_j(W)p - (1-p) \sum_{k=j+1}^{d} c_{k,j} V_k(W) \sum_{s=1}^{k-j} \frac{(-1)^s}{s!} \\
&\times \sum_{\substack{l_1,\ldots,l_s=0 \\ l_1+\cdots+l_s=sd+j-k}}^{d-1} \prod_{r=1}^{s} c_{l_r,d} \mathbf{E}V_{l_r}(\Xi_0),
\end{aligned}
$$

with $c_{k,j} = (k! b_k)/(j! b_j)$, see [WW84]. After dividing by $\mathrm{mes}(W)$, letting $W \uparrow \mathbf{R}^d$ and using the homogeneity property of the functionals involved we get the formulae for the spatial density of V_j,

$$
D_j(\Xi) = (1-p)c_{d,j} \sum_{s=1}^{d-1} \frac{(-1)^s}{s!} \sum_{\substack{l_1,\ldots,l_s=0 \\ l_1+\cdots+l_s=sd+j-k}}^{d-1} \prod_{r=1}^{s} c_{l_r,d} \mathbf{E}V_{l_r}(\Xi_0) \tag{3.39}
$$

for $j = 0, \ldots, d-1$.

Unfortunately, it is impossible to compute the asymptotic variance in the limit theorem for

$$
\mathrm{mes}(W)^{1/2} \left(\frac{\mathbf{E}V_j(\Xi \cap W)}{\mathrm{mes}(W)} - D_j(\Xi) \right)
$$

using the same approach as above. This is explained by the lack of *second-order* integral geometric formulae, which should give decompositions for integrals of V_j^2 or $V_j V_l$ in (3.38).

Extremal properties of the specific connectivity number. The specific connectivity number of an isotropic planar Boolean model is given by (2.14). In the anisotropic planar case the result of Weil [Wei88] gives

$$
\chi_A = (1-p)(\lambda - \lambda^2 \mathsf{A}(\mathbf{E}\Xi_0, -\mathbf{E}\Xi_0)),
$$

where $p = A_A$ is the area fraction, $\mathbf{E}\Xi_0$ is the Aumann expectation of Ξ_0, $-\mathbf{E}\Xi_0 = \{-x : x \in \mathbf{E}\Xi_0\}$, and $\mathsf{A}(\cdot, \cdot)$ is the mixed area [Sch93b, p. 321]. The mixed area is defined via the decomposition

$$
\mathsf{A}(K \oplus K') = \mathsf{A}(K) + \mathsf{A}(K') + 2\mathsf{A}(K, K') \tag{3.40}
$$

for convex sets K and K'.

If $\mathbf{E}\Xi_0$ is central symmetric, then $\mathsf{A}(\mathbf{E}\Xi_0, -\mathbf{E}\Xi_0) = \mathsf{A}(\mathbf{E}\Xi_0, \mathbf{E}\Xi_0) = \mathsf{A}(\mathbf{E}\Xi_0)$, whence

$$
\chi_A = (1-p)(\lambda - \lambda^2 \mathsf{A}(\mathbf{E}\Xi_0)).
$$

The isoperimetric inequality yields an extremal property of the specific connectivity number: namely, for two Boolean models with central symmetric grains and the same p, $\mathbf{EU}(\Xi_0)$ and λ, the isotropic one has the lower χ_A. Roughly speaking, the isotropic version has less 'clumps' and more 'holes' on average, see [Wei88].

Three-dimensional formulae. In the three-dimensional case (in $\mathbf{R}^3$) the formula for χ_A is more complicated than (2.14). For instance, in the isotropic case (3.39) yields

$$\chi_A = (1-p)\left[\lambda - \frac{1}{4\pi}\lambda^2\mathbf{EM}(\Xi_0)\mathbf{ES}(\Xi_0) + \frac{\pi}{96}\lambda^3[\mathbf{ES}(\Xi_0)]^3\right],$$

where $\mathsf{M}(\Xi_0)$ and $\mathsf{S}(\Xi_0)$ are respectively the mean width and the surface area of Ξ_0. The anisotropic case (also for general dimensions) is discussed in [Wei88, Wei90]. The corresponding formulae can be obtained by means of the so-called *Blaschke expectation* of the grain, see p. 54.

3.5   Set-Valued Aggregate Parameters

Convexification

Let us note that most of the numerical aggregate parameters were introduced using the following idea (see [Mat75, pp. 117–119], [SKM95, pp. 16–18, 235–240], [WW84]):

> Take a geometric functional on the family of convex bodies. Then extend this functional for finite unions of convex sets (onto the convex ring). The resulting functional per unit area is the aggregate parameter of interest.

Weil [Wei93b, Wei95] considered also an additive extension of the support function of a convex body onto the convex ring: namely,

$$\bar{h}(F_1 \cup F_2, u) = h(F_1, u) + h(F_2, u) - h(F_1 \cap F_2, u)$$

for convex compact sets F_1 and F_2. Note that $\bar{h}(F_1 \cup F_2, \cdot)$ is *not equal* to the support function of the set $F_1 \cup F_2$. However, $\bar{h}(F_1 \cup F_2, \cdot)$ is also a support function itself, but of another convex set called the *convexification* of $F_1 \cup F_2$. (This is not true for dimensions $d \geq 3$.)

Note that the convexification is different from the convex hull. For any set with a polygonal boundary, the convexification can be obtained by ordering the boundary segments of the set clockwise according to the direction of their outer normals and fitting these segments together to form a closed convex polygon, see Figure 3.6. This procedure was described in [Wei93b] and [Wei95].

Unfortunately, the algorithmisation and implementation of such a procedure to construct the convexification is not easy. In fact, it can be performed effectively only for polygonal grains [Sch92]. Another way to compute the convexification was suggested in [Rat96], see also p. 54.

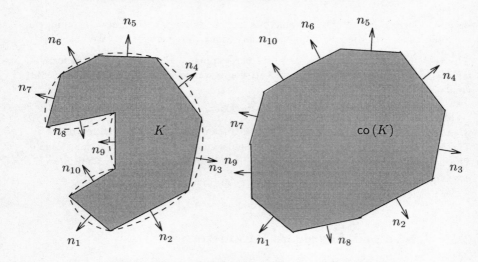

Figure 3.6 Polygon K and its convexification co K.

The additively extended support function $\bar{h}(\Xi \cap W, \cdot)$ is a support function of a set called the convexification of $\Xi \cap W$ and denoted by co $(\Xi \cap W)$. The convexification can be drawn directly through the decomposition of $\Xi \cap W$ into its convex components. To construct the convexification co $(\Xi \cap W)$ it is also possible to use polygonal approximations of $\Xi \cap W$ from inside.

It is possible as well to normalise co $(\Xi \cap W)$ by the area of W. The obtained *specific convexification* co $(\Xi \cap W)/A(W)$ provides an example of a *set-valued* aggregate parameter.

Mean star set

Another example of a set-valued aggregate parameter is the *mean star* set. It is defined as follows. For each $x \notin \Xi$ the set of points visible from x,

$$S_x = \{v \colon [x, v) \cap \Xi = \emptyset\}, \qquad (3.41)$$

is said to be the *star*, see [Ser82, p. 472] and [SKM95, p. 80]. Here $[x, v)$ is the segment that joins x and v with $\{v\}$ excluded. Stationarity of Ξ implies that the distribution of the star-shaped random set S_x does not depend on the choice of the point x outside Ξ. The visibility extent in each direction u is denoted by

$$\zeta_u(S_x) = \sup\{ru \colon ru \in S_x\}. \qquad (3.42)$$

This function is called the *radius-vector* function of the star-shaped random set S_x.

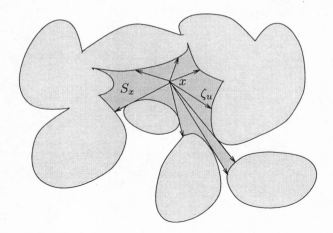

Figure 3.7 Star S_x and radius-vector function $\zeta_u(S_x)$.

The conditional expectations $\zeta_u(\mathcal{S}_\Xi) = \mathbf{E}[\zeta_u(S_x)|x \notin \Xi]$ for all u from the unit circle can be interpreted as the radius-vector function of a closed star-shaped set $\mathcal{S}_\Xi$ called the *radius-vector expectation* of S_x, see [SS94, p. 111].

To estimate the radius-vector expectation let us take a grid, $\mathbb{Z}$, of points on the plane. For each point x from the grid, which does not belong to Ξ, compute the corresponding radius-vector function (3.42). The averages of these radius-vector functions at different points give the radius-vector function of $\mathcal{S}_\Xi$. Therefore, this radius-vector function, $\zeta_u(\mathcal{S}_\Xi)$, is estimated by

$$\hat{\zeta}_{u,W}(\mathcal{S}_\Xi) = (\mathsf{N}(\mathbb{Z} \cap (W \setminus \Xi)))^{-1} \sum_{x \in \mathbb{Z} \cap (W \setminus \Xi)} \zeta_u(S_x).$$

In this way the set $\mathcal{S}_\Xi$ can be estimated. Further visibility properties of the Boolean model were considered in [YZ85]. Some statistical procedures for star-shaped sets were considered in [Gal87].

Steiner compact

Another example of a set-valued parameter is the *Steiner compact* of the fibre process $\partial\Xi$, see Section 3.3. It can be estimated through the rose of intersections of the fibre process $\partial\Xi$, see [MS94b, RS89]. For this, let us take the family of lines $\ell(\beta)$ with varying angle β between the line and the fixed direction (say, of the x-axis). For each line, $\ell(\beta) \cap \partial\Xi$ is a point process on $\ell(\beta)$, see Figure 3.8. Its intensity is denoted by $P_L(\beta)$. The function $P_L(\beta)$, $0 \le \beta < 2\pi$, is called the *rose of intersections* of $\partial\Xi$. It is possible to prove that $P_L(\beta)/2$ is the support function of a convex set $\mathfrak{S}$ called the *Steiner compact* of $\partial\Xi$.

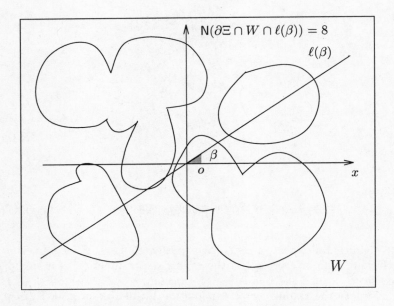

Figure 3.8   Intersection of $\partial\Xi$ with a line.

The practical estimation of $\mathfrak{S}$ is reduced to the estimation of the intensities of point processes that are obtained as linear sections of the boundary $\partial\Xi$. For each $0 \leq \beta < 2\pi$ we estimate $P_L(\beta)$ as the quotient of the number of intersections of $\partial\Xi$ and $\ell(\beta)$ within W and the length of $W \cap \ell(\beta)$,

$$\hat{P}_{L,W}(\beta) = \frac{\mathsf{N}(\partial\Xi \cap W \cap \ell(\beta))}{\mathsf{U}(W \cap \ell(\beta))}. \tag{3.43}$$

The corresponding estimator of $\mathfrak{S}$ is given by the maximal convex set with the support function lower than $\hat{P}_{L,W}(\beta)/2$,

$$\hat{\mathfrak{S}}_W = \left\{ v\colon \langle v, u \rangle \leq \hat{P}_{L,W}(\beta)/2,\ u = (\cos\beta, \sin\beta),\ 0 \leq \beta < 2\pi \right\}. \tag{3.44}$$

In practice, one takes only a finite number of angles β. Then the resulting estimator is a polygon, which approximates $\mathfrak{S}$ as the number of angles increases, see [RS89].

One half of the perimeter of the Steiner compact is equal to the intensity of the fibre process $\partial\Xi$ [SKM95, p. 289]. The latter is equal to the specific boundary length, i.e.

$$L_A = \frac{1}{2}\mathsf{U}(\mathfrak{S}) = \frac{1}{4}\int\limits_{0}^{2\pi} P_L(\beta)d\beta. \qquad (3.45)$$

This formula allows us to estimate also the specific boundary length of the Boolean model.

Notes to Section 3.5

Surface measures and convexifications. Let us consider a planar regular closed set F from the convex ring. The regular closedness property ($F = \overline{\text{Int } F}$) implies the existence of the outer normal $\mathbf{n}_F(x)$ for almost all points $x \in \partial F$. Then

$$S_1(F;\Gamma) = \mathcal{H}^1(\{x \in \partial F : \mathbf{n}_F(x) \in \Gamma\}) \qquad (3.46)$$

defines the *direction measure* of F (or surface area measure of order 1), see also p. 42. Here $\mathcal{H}^1$ denotes the 1-dimensional Hausdorff measure (or the boundary length). A relationship between the direction measure and the convexification is given by the identity

$$S_1(F, \cdot) = S_1(\text{co}(F), \cdot)$$

which is valid for any set F from the convex ring [Wei93b]. Thus, $\text{co}(F)$ is the unique convex set with the same direction measure as F. In fact, the direction measure determines the distribution of the (oriented) outer normal vector to F. As was pointed out by Weil [Wei93b], it is possible to extend this construction for planar regular closed sets *not* belonging to the convex ring.

In turn, the direction measure is related to the mixed area by

$$\mathsf{A}(F, K) = \frac{1}{2}\int\limits_{\mathsf{S}^{d-1}} h(K, u)S_1(F, du). \qquad (3.47)$$

This relation suggests the following way to build the convexification, $\text{co}(F)$, see [Rat96]. First, estimate the mixed area, for example, by

$$\mathsf{A}_\varepsilon(F, K) = \frac{\mathsf{A}(F \oplus \varepsilon K) - \mathsf{A}(F)}{2\varepsilon}$$

for a suitably small ε and K belonging to a certain family of sets (e.g., sectors, three-point sets or equilateral triangles) [Rat96]. The solution of the integral equation (3.47) for K running through the chosen family of compact sets yields an estimator of the direction measure of F. This integral equation can be solved using Fourier expansions and a discretised version of the integral in (3.47). Finally, the direction measure allows us to reconstruct the corresponding convexification up to a translation.

The Blaschke expectation. This expectation is defined as follows via expected surface area measures. First, consider the surface area measure $S_{d-1}(\Xi_0; \Gamma)$ defined by (3.31), see also [Sch93b, p. 203]. Then the Blaschke expectation, $\mathbf{E}_B\Xi_0$, of Ξ_0 is defined by the identity

$$\mathbf{E}S_{d-1}(\Xi_0;\Gamma) = S_{d-1}(\mathbf{E}_B\Xi_0;\Gamma)$$

for all Borel sets $\Gamma \subset \mathbf{S}^{d-1}$. Thus, the Blaschke expectation of a set coincides with the Blaschke expectation of its convexification,

$$\mathbf{E}_B\Xi_0 = \mathbf{E}_B\mathrm{co}\,(\Xi_0)\,.$$

Note that the Blaschke and the Aumann expectation of a convex *planar* random compact set coincide. However, this is not true in higher dimensions.

The rose of directions and the rose of intersections. The direction measure of a planar set F (see p. 53) gives the distribution of the oriented outer normal to ∂F. Sometimes also the tangent directions are of interest. The corresponding measure $R(\cdot)$ on the unit sphere (distribution of the tangent vector) is said to be the *rose of directions* of F. This concept is useful when handling fibre processes, where there is no natural way to define outer directions. If F is a planar set, then, rotated by $\pi/2$, $R(\cdot)$ gives the distribution of the non-oriented normal vector.

In the case of fibre processes another interesting rose is the *rose of intersections*, $P_L(\cdot)$, see p. 52. These two roses are related by the following formula:

$$P_L(\beta) = L_A \int\limits_{(0,\pi]} |\sin(\alpha - \beta)| R(d\alpha)\,.$$

This equation is well known in convex geometry, see [SW83]. Directional characteristics of fibre processes were discussed also in [MS80, RS89, RS92, SMP80].

Let us outline related estimation problems in statistics of the Boolean model, see [MS94b]. For this, suppose that Ξ is an anisotropic Boolean model with a convex typical grain. The intersection $\Xi \cap \ell(\beta)$ is a one-dimensional Boolean model with a segment as typical grain. Its random length, denoted by $\xi(\beta)$, is called the length of the random intercept of the grain Ξ_0 in the direction $\ell(\beta)$ [Mat75, p. 87]. Note that the length fraction of $\Xi \cap \ell(\beta)$ cannot serve as a directional characteristic of Ξ, since it is constant for all β. On the other hand, $m(\beta) = \mathbf{E}\xi(\beta)$ and the intensity $\lambda(\beta)$ of $\Xi \cap \ell(\beta)$ depend on β in the anisotropic case, and, therefore, can be used as directional characteristics. The function $m(\beta)$ is called the rose of mean chord length [MS94b]. It is easy to prove that

$$\begin{aligned} P_L(\beta) &= 2\lambda(\beta)(1-p)\,, \\ \lambda(\beta)m(\beta) &= \lambda\mathbf{E}A(\Xi_0)\,, \end{aligned}$$

and also

$$\int\limits_0^\pi \frac{d\beta}{m(\beta)} = \frac{\pi}{\mathbf{E}\sigma}\,,$$

where $\mathbf{E}\sigma$ is the mean length of the random chord in Ξ_0 defined by an isotropic uniformly random line [San76, p. 30]. Estimators for different roses are considered in [MS94b]. In particular, their uniformity with respect to β has been proved.

Steiner compacts and associated zonoids. An even measure ν on $\mathbf{S}^{d-1}$ is related to a convex body, Π_ν, called the *associated zonoid* of ν. This body Π_ν is defined by its support function

$$h(\Pi_\nu, u) = \int_{\mathbf{S}^{d-1}} |\langle u, v \rangle| \nu(dv).$$

If ν is the non-oriented distribution of the normal vector of the Boolean model Ξ, then the corresponding associated zonoid is equal (up to a constant determined by the intensity) to the Steiner compact. The concept of associated zonoid allows us to invoke integral geometric methods to solve extremal problems in stochastic geometry, see, e.g., [Wei88, Wie86, Wie89].

Uncovered components. The complement of the Boolean model can be explored through the corresponding star sets. For high-intensity Boolean models it is possible as well to consider all connected vacant components. It was proven in [Hal85a, Mol96a] that a scaled elementary connected vacant component converges in distribution to the typical polyhedron generated by an (in general anisotropic) Poisson network of hyperplanes driven by the expected surface measure of the grain.

A simple result of this kind can be explained as follows, see [Mol96a]. Let $\Xi(n)$ be the Boolean model of intensity λ_n with a convex regular closed typical grain Ξ_0 which distribution does not depend on n, and let Y_n be the connected component of $\mathbf{R}^d \setminus \Xi(n)$ containing the origin. Then

$$\mathbf{P}\left\{ K \subset \lambda_n Y_n \,|\, o \notin \Xi(n) \right\} \to \exp\left\{ -\int_{\mathbf{S}^{d-1}} h(K \cup \{o\}, u) \mathbf{E} S_{d-1}(\Xi_0; du) \right\}, \quad K \in \mathcal{K},$$

as $n \to \infty$ and $\lambda_n \uparrow \infty$. The limiting distribution corresponds to the Poisson polyhedron in $\mathbf{R}^d$ generated by a network of hyperplanes in $\mathbf{R}^d$. The normal vectors to the hyperplanes from the network have the distribution ν on $\mathbf{S}^{d-1}$, such that the corresponding associated zonoid is equal to the projection body [GW93a] of the Blaschke expectation, $\mathbf{E}_B \Xi_0$. This result was used in [Mol96a] to suggest an estimation procedure for Boolean models with high intensities.

4

Estimation of Functional Aggregate Parameters

It should be noted that the set-valued aggregate parameters considered in Section 3.5 are related to some functions. For example, the convexification and the Steiner compact can be represented through their support functions. In this chapter we will consider other functional aggregate parameters which do not admit direct set-valued reformulations.

4.1   Capacity Functional

The distribution of the Boolean model Ξ is determined by the corresponding capacity functional T_Ξ given by (2.21), which is the most comprehensive aggregate parameter. The estimation of T_Ξ from a realisation of Ξ is discussed below.

Let K be a compact set. For each point x we can check if $K + x$ hits Ξ. Since Ξ is ergodic, the area fraction of all $x \in \mathbf{R}^2$ satisfying $(K + x) \cap \Xi \neq \emptyset$ gives the corresponding hitting probability,

$$\mathbf{P}\left\{(K + x) \cap \Xi \neq \emptyset\right\} = \mathbf{P}\left\{\Xi \cap K \neq \emptyset\right\} = T_\Xi(K).$$

In fact, $(K + x) \cap \Xi \neq \emptyset$ if and only if x belongs to the set $\Xi \oplus \check{K}$. Therefore, the value $T_\Xi(K)$ is equal to the volume fraction of the dilated set $\Xi \oplus \check{K}$.

If Ξ is observed inside window W only, then it is not possible to check whether Ξ and $K + x$ have a non-empty intersection if $K + x$ stretches beyond W, see Figure 4.1.

This problem can be handled by using the *minus-sampling* approach which means that only the information contained sufficiently 'deep' in the window

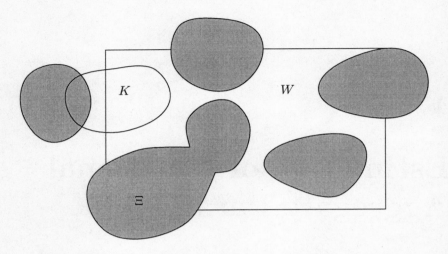

Figure 4.1   Edge effect: K hits Ξ but does not hit $\Xi \cap W$.

is used. For each compact set K, the corresponding 'minus-window' is given by

$$W \ominus K = \{x: \ x + K \subset W\}$$

and is called the *erosion* of W by K, see [Ser82, Hei94a]. Note that $x \in W \ominus K$ implies $x + K' \subset W$ for each $K' \subset K$. This is a particular form of the *local knowledge* principle of Serra [Ser82]:

> From knowledge of $\Xi \cap W$ the dilation of Ξ can be computed inside the erosion of W.

Then an estimator of T is given by

$$\hat{T}_{\Xi,W}(K) = \frac{\mathsf{A}((\Xi \oplus \check{K}) \cap (W \ominus K))}{\mathsf{A}(W \ominus K)}, \quad K \in \mathcal{K}. \tag{4.1}$$

Further, we refer to $\hat{T}_{\Xi,W}(K)$ as the *empirical capacity functional* of Ξ. The computational technique was discussed in [Rip86]. Since $T_{\Xi}(K)$ coincides with the area fraction of the dilated Boolean model $\Xi \oplus \check{K}$, (4.1) is a *strong consistent* estimator of $T_{\Xi}(K)$, i.e.

$$\hat{T}_{\Xi,W}(K) \to T_{\Xi}(K) \quad \text{a.s. as} \quad W \uparrow \mathbf{R}^2 \,.$$

In the simplest particular case K is a singleton $\{x\}$. Because of stationarity, the value of $T_{\Xi}(\{x\})$ on such a set does not depend on x, so that (4.1) yields the empirical area fraction of Ξ.

If the typical grain coincides almost surely with the closure of its interior, then estimator (4.1) is *uniformly* strong consistent, i.e. for each compact set K_0,

$$\sup_{K \subset K_0,\, K \in \mathcal{K}} |\hat{T}_{\Xi,W}(K) - T_\Xi(K)| \to 0 \quad \text{a.s. as} \quad W \uparrow \mathbf{R}^2, \tag{4.2}$$

see [Mol89] and Theorem 4.5. This result is also valid for higher dimensions. The technique of Mase [Mas82] can be applied to investigate asymptotic properties of estimator (4.1), although the computations are very complicated even for simple typical grains. For example, $\mathsf{A}(W)^{1/2}(\hat{T}_{\Xi,W}(K) - T_\Xi(K))$ is asymptotically normal as $W \uparrow \mathbf{R}^2$ with the asymptotic variance given by

$$
\begin{aligned}
\sigma_T^2(K) &= \lim_{W \uparrow \mathbf{R}^2} \mathsf{A}(W) \operatorname{Var} \hat{T}_{\Xi,W}(K) \\
&= \int_{\mathbf{R}^2} \left(2T_\Xi(K) - T_\Xi(K \cup (K+v)) - T(K)^2 \right) dv \\
&= \int_{\mathbf{R}^2} (C_{\Xi \oplus \check{K}}(v) - p_{\Xi \oplus \check{K}})^2 dv,
\end{aligned}
\tag{4.3}
$$

where $C_{\Xi \oplus \check{K}}(\cdot)$ (resp. $p_{\Xi \oplus \check{K}}$) is the covariance function (resp. area or volume fraction) of the Boolean model $\Xi \oplus \check{K}$. Unfortunately, (4.3) involves the values of the capacity functional T_Ξ. Mase [Mas82] suggested computing $\sigma_T^2(K)$ via an empirical estimator of the covariance function for the Boolean model $\Xi \oplus \check{K}$.

Sometimes estimators of the logarithmic transform of T_Ξ are needed. Put

$$\Psi_\Xi(K) = -\log(1 - T_\Xi(K)), \tag{4.4}$$

and define its empirical counterpart by

$$\hat{\Psi}_{\Xi,W}(K) = -\log(1 - \hat{T}_{\Xi,W}(K)). \tag{4.5}$$

Then, for each K_0 with $T(K_0) < 1$,

$$\sup_{K \subset K_0,\, K \in \mathcal{K}} |\hat{\Psi}_{\Xi,W}(K) - \Psi_\Xi(K)| \to 0 \quad \text{a.s. as} \quad W \uparrow \mathbf{R}^2. \tag{4.6}$$

Furthermore, the random variable

$$\mathsf{A}(W)^{1/2} \left(\hat{\Psi}_{\Xi,W}(K) - \Psi_\Xi(K) \right)$$

converges in distribution to the Gaussian random vector with zero mean and the variance

$$\sigma_\Psi^2(K) = \sigma_T^2(K)(1 - T_\Xi(K))^{-2}. \tag{4.7}$$

Notes to Section 4.1

Empirical capacities for iid random sets. Let $A, A_1, A_2, \ldots$ be independent identically distributed random closed sets of general nature (stationarity is not assumed). The corresponding empirical capacity functional is then given by

$$T_n^*(K) = \frac{1}{n} \sum_{i=1}^{n} \mathbf{1}_{A_i \cap K \neq \emptyset} \, .$$

This functional is a particular case of the empirical measure on the space $\mathcal{F}$ of closed sets. If $A = \{\xi\}$ is a singleton, then $T_n^*(\cdot)$ is the empirical distribution of the random vector ξ.

The strong law of large numbers yields the almost sure pointwise convergence of $T_n^*(K)$ to the capacity functional $T_A(K) = \mathbf{P}\{A \cap K \neq \emptyset\}$. However, the complicated geometrical structure of the space $\mathcal{F}$ causes difficulties when investigating the uniform convergence [Mol87, Mol90a].

The random set A is said to satisfy the *Glivenko–Cantelli* theorem with respect to a class $\mathfrak{M} \subset \mathcal{K}$ (notation $A \in \mathsf{GC}(\mathfrak{M})$) if, for any compact set $K_0 \in \mathcal{K}$,

$$\sup_{K \in \mathfrak{M}, \, K \subset K_0} |T_n^*(K) - T_A(K)| \to 0 \quad \text{a.s. as} \quad n \to \infty. \tag{4.8}$$

It is possible to build simple examples of random sets that do *not* satisfy the Glivenko–Cantelli theorem even with respect to a simple $\mathfrak{M}$.

EXAMPLE 4.1 Let $\mathfrak{M} = \{\{x\}: x \in \mathbf{R}^1\}$ be the class of all singletons on the line and let $K_0 = [0, 1]$. Furthermore, let M be a nowhere-dense subset of K_0 with a positive Lebesgue measure. Finally, put $A = M + \xi$ for a Gaussian random variable ξ. Since K_0 is a set of the second Baire category, there exists a point $x_n \in K_0 \setminus \cup_{i=1}^{n} A_i$ for every $n \geq 1$ and all possible realisations $A_1, \ldots, A_n$ of the random closed set A. Hence, $T_n^*(\{x_n\}) = 0$, $n \geq 1$, while

$$T(\{x_n\}) \geq \inf_{x \in [0,1]} T(\{x\}) \geq \varepsilon > 0 \, .$$

Therefore, (4.8) does not hold. One can use similar arguments for the Boolean model with the typical grain $\Xi_0 = M$.

EXAMPLE 4.2 Let $\mathfrak{M} = \{\{x_n\}, \, n \geq 1\}$ be a class containing a countable collection of distinct singletons in $[0, 1]$. Passing to subclasses if necessary, we can assume that $x_n \to x_0$ as $n \to \infty$, and $\{x_0\} \in \mathfrak{M}$. Let $\{\xi_n, \, n \geq 1\}$ be a sequence of independent identically distributed random variables taking the values 0 and 1 with probability 1/2. Define $A = \{x_n: n \geq 1, \, \xi_n = 1\} \cup \{x_0\}$. It follows from the construction of the random closed set A that, for all its independent realisations $A_1, \ldots, A_n$, there exists a point $x_{n(\omega)}$ such that $x_{n(\omega)} \notin \cup_{i=1}^{n} A_i$ almost surely. Hence, $T_n^*(\{x_{n(\omega)}\}) = 0$. Furthermore, $T(\{x_{n(\omega)}\}) = \mathbf{P}\{\xi_i = 1\} = 1/2$, i.e. (4.8) does not hold. Thus, $A \notin \mathsf{GC}(\mathfrak{M})$.

Remember that random closed set A is called almost surely *regular closed* if it almost surely coincides with the closure of its interior, i.e. $A = \overline{\operatorname{Int} A}$ a.s. Furthermore, A is said to be *a.s. continuous* [Mat75] if $\mathbf{P}\{x \in \partial A\} = 0$ for any point $x \in \mathbf{R}^d$.

Theorem 4.3 (see [Mol87]) *Let $\mathfrak{M}$ be a class of compact sets that is closed in the Hausdorff metric. If random closed set A satisfies the conditions:*

A1 *A is almost surely regular closed;*

A2 *for any $K \in \mathfrak{M}$, $T_A(K) = T_{\operatorname{Int} A}(K)$, i.e. $T_A(K) = \mathbf{P}\{\operatorname{Int} A \cap K \neq \emptyset\}$;*

then $A \in \operatorname{GC}(\mathfrak{M})$. Conditions **A1** *and* **A2** *are also necessary if $\mathfrak{M} = \mathcal{K}$ and A is a.s. continuous.*

Conditions for the class $\mathfrak{M}$ to be *universal* (this means that *each* random set satisfies the Glivenko–Cantelli theorem with respect to $\mathfrak{M}$) can be found in [Mol90a].

Empirical capacities for stationary random sets. If X is a stationary ergodic random closed set, then its empirical capacity functionals can be estimated by a single realisation of X. However, this does not affect the uniformity conditions for empirical capacities [Mol89]. It is possible as well to construct pathological examples of stationary random sets and even Boolean models that do not satisfy the Glivenko–Cantelli theorem with respect to a simple $\mathfrak{M}$. For this, it is possible to take random closed sets from Examples 4.1 and 4.2 as their typical grains.

The capacity functional for stationary random closed sets was estimated in [Mol89, Mol92, Mol94a] as follows. Pick a compact set K_0 such that $K \subset K_0$ for all sets K of interest. Then an estimator of T is given by

$$\hat{T}^0_{X,W}(K) = \frac{\operatorname{mes}((X \oplus \check{K}) \cap (W \ominus K_0))}{\operatorname{mes}(W \ominus K_0)}. \tag{4.9}$$

Note that the estimator (4.1) explores better the information contained within the window W. In contrast to (4.1) the estimator $\hat{T}^0_{X,W}(K)$ uses one large 'minus-window' for all $K \subset K_0$. The following result is valid for the estimator (4.9).

Theorem 4.4 (see [Mol89]) *Let X be an ergodic stationary random closed set. Then $X \in \operatorname{GC}(\mathcal{K})$, i.e.*

$$\sup_{K \in \mathcal{K},\, K \subset K_0} |\hat{T}^0_{X,W}(K) - T_X(K)| \to 0 \quad a.s. \ as \quad W \uparrow \mathbf{R}^d \tag{4.10}$$

for each compact set K_0 if (and only if for a.s. continuous X) the random set X is almost surely regular closed.

The proof of Theorem 4.4 is based on the fact that $\mathbf{P}\{X \cap K \neq \emptyset,\ \operatorname{Int} X \cap K = \emptyset\}$ vanishes for each stationary almost surely regular closed random set X and every $K \in \mathcal{K}$. Fortunately, the same result is valid for the estimator (4.1).

Theorem 4.5 *If X is an ergodic stationary a.s. regular closed random set, then (4.2) holds for each fixed compact set K_0.*

PROOF. Clearly, for every $K \subset K_0$,

$$|\hat{T}_{X,W}(K) - \hat{T}^0_{X,W}(K)|$$
$$\leq \text{mes}(W \ominus K_0)^{-2} \Big[\text{mes}(W \ominus K) \big| \text{mes}((X \oplus \check{K}) \cap (W \ominus K_0))$$
$$- \text{mes}((X \oplus \check{K}) \cap (W \ominus K)) \big|$$
$$+ \text{mes}((X \cap \check{K}) \cap (W \ominus K)) \big| \text{mes}(W \ominus K) - \text{mes}(W \ominus K_0) \big| \Big]$$
$$\leq \text{mes}(W \ominus K_0)^{-2} \text{mes}(W \ominus K) 2 \Big(\text{mes}(W \ominus K) - \text{mes}(W \ominus K_0) \Big)$$
$$\leq \text{mes}(W \ominus K_0)^{-2} \text{mes}(W) 2 \Big(\text{mes}(W) - \text{mes}(W \ominus K_0) \Big).$$

By assumption (see p. 9), $W \uparrow \mathbf{R}^d$ in such a way that $\text{mes}(W)/\text{mes}(W \ominus K_0) \to 1$. Thus,

$$\sup_{K \in \mathcal{K},\ K \subset K_0} |\hat{T}_{X,W}(K) - \hat{T}^0_{X,W}(K)| \to 0 \quad \text{a.s. as} \quad W \uparrow \mathbf{R}^d,$$

and (4.2) follows from Theorem 4.4. $\square$

Limiting variance. Using (2.21), formula (4.3) can be transformed as follows (for general dimension d):

$$\sigma^2_T(K)$$
$$= \exp\{-2\lambda \mathbf{E} \text{mes}(\Xi_0 \oplus \check{K})\} \int_{\mathbf{R}^d} \left(\exp\{\lambda \mathbf{E} \text{mes}((\Xi_0 \oplus \check{K}) \cap (\Xi_0 \oplus \check{K} - v))\} - 1 \right) dv.$$

Note that

$$\sigma^2_T(K) \leq \exp\{-\lambda \mathbf{E} \text{mes}(\Xi_0 \oplus \check{K})\} \lambda \int_{\mathbf{R}^d} \mathbf{E} \text{mes}((\Xi_0 \oplus \check{K}) \cap (\Xi_0 \oplus \check{K} - v)) dv$$
$$= (1 - T_\Xi(K)) \lambda \mathbf{E}[\text{mes}(\Xi_0 \oplus \check{K})]^2.$$

Thus, $\sigma^2_T(K) < \infty$ if

$$\mathbf{E}[\text{mes}(\Xi_0 \oplus B_r)]^2 = \mathbf{E} \text{mes}(\Xi^r_0)^2 < \infty$$

for all $r > 0$. If Ξ_0 is a.s. convex, then it suffices to assume that $\mathbf{E} V_j(\Xi_0)^2 < \infty$ for all $j = 1, \ldots, d$, because of the Steiner formula (2.26).

Functional limit theorems. We will deal here with the empirical capacity functional of the Boolean model Ξ. It should be noted that similar results are valid for any stationary mixing random closed set [Hei92b] and for empirical capacities constructed by its independent realisations as well [Mol90b].

It follows from the mixing condition [Hei92b] that the random vector

$$A(W)^{1/2} \Big(\hat{T}_{\Xi,W}(K_1) - T_\Xi(K_1),\ \hat{T}_{\Xi,W}(K_2) - T_\Xi(K_2) \Big)$$

converges in distribution as $W \uparrow \mathbf{R}^d$ to the Gaussian vector with zero mean and the covariance matrix

$$\begin{pmatrix} \sigma_T^2(K_1) & \sigma_T(K_1, K_2) \\ \sigma_T(K_1, K_2) & \sigma_T^2(K_2) \end{pmatrix}$$

with

$$\sigma_T(K_1, K_2) = \int_{\mathbf{R}^d} \Big(T_\Xi(K_1) + T_\Xi(K_2) - T_\Xi(K_1 \cup (K_2 + v)) - T_\Xi(K_1)T_\Xi(K_2) \Big) dv.$$

Clearly, $\sigma_T^2(K) = \sigma_T(K, K)$. Furthermore, finite-dimensional distributions of the random field

$$\zeta_W(K) = \mathsf{A}(W)^{1/2} \Big(\hat{T}_{\Xi, W}(K) - T_\Xi(K) \Big), \quad K \in \mathcal{K},$$

converge to the finite-dimensional distributions of a centred Gaussian random process $\zeta(K)$, $K \in \mathcal{K}$, with the covariance

$$\mathbf{E}[\zeta(K_1)\zeta(K_2)] = \sigma_T(K_1, K_2). \tag{4.11}$$

However, to ensure the *functional* convergence of $\zeta_W(K)$, $K \in \mathfrak{M}$, the class $\mathfrak{M}$ and the functional $T_\Xi(\cdot)$ must satisfy additional conditions, which are similar to those used in the theory of empirical measures and set-indexed random functions [Dud84, Pyk83].

Theorem 4.6 (see [Mol90b]) *Suppose that $K \subset K_0$ for all $K \in \mathfrak{M}$ and that*

C1 $\log \nu(\varepsilon) = \mathcal{O}(\varepsilon^{-\beta})$ *for some $\beta \in (0,1)$, where $\nu(\varepsilon)$ is the cardinality of the minimum ε-net of $\mathfrak{M}$ in the Hausdorff metric (this ε-net is a finite set such that its ε-neighbourhood in the Hausdorff metric covers $\mathfrak{M}$);*

C2 *there exists a $\gamma > \beta$, such that*

$$\sup_{\rho_H(K_1, K_2) < \varepsilon, \ K_1, K_2 \in \mathfrak{M}} |T_\Xi(K_1) - T_\Xi(K_2)| = \mathcal{O}(\varepsilon^\gamma).$$

Then $\zeta_W(\cdot)$ converges weakly in the uniform metric to the Gaussian random field $\zeta(\cdot)$ on $\mathfrak{M}$ with the covariance (4.11).

The statement of Theorem 4.6 means that every continuous in the uniform metric functional of ζ_W converges in distribution to its value on ζ. Similar results are valid for the functional Ψ_Ξ and its empirical estimator $\hat{\Psi}_{\Xi, W}$ with the limiting covariance given by

$$\sigma_\Psi(K_1, K_2) = (1 - T_\Xi(K_1))^{-1}(1 - T_\Xi(K_2))^{-1}\sigma_T(K_1, K_2). \tag{4.12}$$

Non-stationary Boolean model. The capacity functional of a non-stationary Boolean model Ξ must be estimated by independent realisations of Ξ. The resulting empirical capacity converges uniformly if the grain is a.s. regular closed and the intensity measure Λ of the (non-stationary) Poisson process Π_Λ of germs satisfies $\Lambda(G) = \Lambda(\bar{G})$ for any open set G, see [Mol91a].

4.2 Covariance

The *exponential* covariance function (2.19) can be estimated using (2.20) by means of estimates of numerical aggregate parameters, L_A and p. In the general case it is possible to express the covariance function through the capacity functional on the family of two-point sets:

$$C(v) = 2p - T_\Xi(\{o, v\}).$$

Then the replacement of p by the empirical area fraction $\hat{p}_W$ and $T_\Xi(\cdot)$ by the empirical capacity functional $\hat{T}_{\Xi,W}(\cdot)$ yields an estimator $\hat{C}_W(v)$ of the covariance function. The evaluations can be simplified by noticing that $C(v)$ is equal to the area fraction of the two-point erosion

$$\Xi \ominus \{o, v\} = \{x\colon \{x, x + v\} \subset \Xi\}$$

of Ξ. Therefore, an unbiased estimator of $C(v)$ is given by

$$\hat{C}_W(v) = \frac{\mathrm{mes}((\Xi \cap W) \ominus \{o, v\})}{\mathrm{mes}(W \ominus \{o, v\})}. \tag{4.13}$$

Sometimes an estimator $\hat{q}_W(v)$ of the function $q(v)$ from (2.18) is necessary. It can be obtained by replacing in (2.18) all ingredients by their empirical counterparts.

It follows from (4.2) that the estimators $\hat{C}_W(v)$ and $\hat{q}_W(v)$ are *uniformly strong consistent*, e.g.,

$$\sup_{v \in K_0} |\hat{q}_W(v) - q(v)| \to 0 \quad \text{a.s. as} \quad W \uparrow \mathbf{R}^2$$

for each compact set K_0. However, the construction of *functional* confidence intervals for the covariance is not easy. Note that the relative fluctuations of $\hat{C}_W(v)$ and $\hat{q}_W(v)$ are especially noticeable for v with a large norm, i.e. when $q(v)$ is close to 1 (or $C(v)$ is close to p^2).

If the Boolean model Ξ is isotropic, then it is advisable to estimate the covariance using averages for all directions: namely, in this case $C(v)$ is estimated by

$$\hat{C}_W^{(i)}(\|v\|) = \frac{1}{2\pi} \int_{S^1} \hat{C}_W(\|v\|u)\,du.$$

For higher dimensions we can average $\hat{C}_W(\|v\|u)$ for u from the unit sphere $\mathbf{S}^{d-1}$.

Notes to Section 4.2

Three-point covariance. Sometimes the three-point analogue of the covariance can be of use. This function is defined by

$$C(v_1, v_2) = \mathbf{P} \left\{ \{o, v_1, v_2\} \subset \Xi \right\} .$$

Clearly, it can be estimated through erosions of Ξ by three-point sets

$$\Xi \ominus \{o, v_1, v_2\} = \{x \colon \{x, x + v_1, x + v_2\} \subset \Xi \} .$$

The corresponding empirical estimator is uniformly strong consistent for all v_1, v_2 from an arbitrary fixed compact set.

4.3   Contact Distribution Functions

Let B be a convex set, containing the origin. Then $rB \subset r_1 B$ for all $s < s_1$. Therefore, the area fraction of $\Xi \oplus rB$ increases as r gets larger, whence the function $T_\Xi(rB)$, $r \geq 0$, is non-decreasing. After a simple transformation this non-decreasing function becomes a distribution function. Put

$$H_B(r) = 1 - \frac{1 - T_\Xi(rB)}{1 - p}, \quad r \geq 0. \tag{4.14}$$

Then $H_B(0) = 0$, $H_B(\infty) = 1$ and the function H_B is right-continuous, i.e. $H_B(r)$ is a distribution function of a positive random variable that has the following probabilistic interpretation. Introduce the *contact distance*, ζ_B, by

$$\zeta_B = \inf \{r \geq 0 \colon rB \cap \Xi \neq \emptyset\} , \tag{4.15}$$

see Figure 4.2. Then $H_B(r)$ is the distribution function of the random variable ζ_B under the condition that the origin is not covered by Ξ, i.e.

$$H_B(r) = \mathbf{P} \left\{ \zeta_B < r | o \notin \Xi \right\} .$$

The function $H_B(r)$ is said to be the *contact distribution function* and B is said to be the *structuring element*, see [Cre91, Ser82] and [SKM95, Chapter 3]. It follows from (2.21) that

$$H_B(r) = 1 - \exp \left\{ - \lambda \mathbf{E} \big[\mathsf{A}(\Xi_0 \oplus r\breve{B}) - \mathsf{A}(\Xi_0) \big] \right\}. \tag{4.16}$$

Three particular cases are worth mentioning.

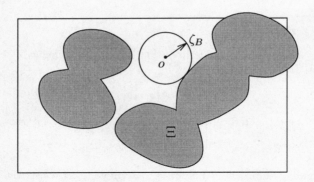

Figure 4.2 ζ_B for B equal to the unit disk

- B is the unit disk on the plane (or the unit ball for higher dimensions). Then $H_B(r)$ is denoted by $H_o(r)$ and said to be the *spherical* contact distribution function. The contact distance, ζ_B from (4.15) is distributed as the distance from the origin to the nearest point of Ξ.
- B is the unit square (cube). Then $H_B(r)$ is denoted by $H_\square(r)$ and said to be the *quadratic* contact distribution function. It serves as a replacement of the spherical contact distribution function on the discretised computer screen, especially for small values of r.
- If B is the segment $[0, u]$ for a point u on the unit circle (sphere), then $H_B(r) = H_u(r)$ is the *linear* contact distribution function. In contrast to the two previous cases it can be used (with varying u) to explore the anisotropy of the Boolean model Ξ.

The contact distribution function can be estimated from the estimators of the capacity functional and the area fraction: namely, put

$$\hat{H}_{B,W}(r) = 1 - \frac{1 - \hat{T}_{\Xi,W}(rB)}{1 - \hat{p}_W}, \qquad (4.17)$$

see [Hei93]. As above,

$$\hat{T}_{\Xi,W}(rB) = \frac{\mathrm{mes}((\Xi \oplus r\breve{B}) \cap (W \ominus rB))}{\mathrm{mes}(W \ominus rB)} \qquad (4.18)$$

is the minus-sampling estimator of the capacity functional, and $\hat{p}_W = \mathrm{mes}(\Xi \cap W)/\mathrm{mes}(W)$ is the empirical volume fraction. The estimator (4.17) is uniformly strong consistent:

$$\sup_{r \geq 0} |\hat{H}_{B,W}(r) - H_B(r)| \to 0 \quad \text{a.s. as} \quad W \uparrow \mathbf{R}^2 .$$

The proof follows the proof of the classical Glivenko–Cantelli theorem (see, e.g., [Loè63, p. 20]), since $H_B(r)$ is a distribution function for any star-shaped set B. Moreover, (4.2) yields the uniformity with respect to all compact structuring elements B that are subsets of a fixed compact set K_0. For instance, the empirical linear contact distribution function is uniformly strong consistent not only with respect to r, but also with respect to $u \in K_0$. Limit theorems for contact distribution functions [Bad80, Hei93] can be (at least theoretically) used to construct confidence intervals.

Unfortunately, the estimator $\hat{H}_{B,W}(r)$ may not be monotone. This can be explained by the changing denominator in (4.18) due to the minus-sampling edge-correction. This problem can be overcome by using the estimator $\hat{T}^0_{\Xi,W}(rB)$ instead of $\hat{T}_{\Xi,W}(rB)$ with $0 < r \le r_0$ for a fixed r_0 and $K_0 = r_0 B$, see (4.9). However, such a reduction of the window implies further loss of information. Below (see p. 68) we consider another (so-called Kaplan–Meier) estimator, which respects the monotonicity of the contact distribution function.

Sometimes we need the logarithmic transformation of $H_B(r)$ given by

$$H^l_B(r) = -\log(1 - H_B(r))\,. \tag{4.19}$$

This function is said to be the *logarithmic* (spherical, quadratic, linear, etc.) contact distribution function. It follows from (4.16) that

$$H^l_B(r) = \lambda \mathbf{E}\big[\mathrm{A}(\Xi_0 \oplus r\breve{B}) - \mathrm{A}(\Xi_0)\big]\,. \tag{4.20}$$

Plots of contact distribution functions reveal some features of the Boolean model and they may be used for graphical tests, see Section 9.1. Later on it will be shown how to relate the individual parameters of the Boolean model to the corresponding contact distribution functions, see Section 5.1. It is also possible to use plots of contact distribution functions to compare images without addressing the Boolean model assumption, see [AM93, MA91, Rip86].

Notes to Section 4.3

General structuring elements. Contact distribution functions can be introduced for any stationary random closed set Z and for a general star-shaped structuring element B. It has been pointed out in [Hei93] that $H_B(\cdot)$ becomes a distribution function if either the interior of B is not empty or the set Z is mixing, see (2.24). Otherwise $H_B(\infty)$ can be strictly less than 1.

EXAMPLE **4.7** (see [Hei93]) Consider the stationary (and ergodic, see (2.25)) random closed set

$$Z = \bigcup_{z \in \mathbb{Z}^2} B_\delta(z) + \xi\,,$$

where $\mathbb{Z}^2$ is the grid of integers in $\mathbf{R}^2$ and ξ is uniformly distributed on the unit square $[0,1]^2$. Then for the unit segment $B = [o, e_0]$ with the end-point $e_0 = (1,0)$ we get $\mathbf{P}\{Z \cap rB = \emptyset\} = 1 - 2\delta$ for all $r \geq 1$, whence

$$H_B(\infty) = \frac{2\delta - \pi\delta^2}{1 - \pi\delta^2} < 1.$$

Functional limit theorems for contact distribution functions. The empirical contact distribution function $\hat{H}_{B,W}(r)$ given by (4.17) is a strong consistent estimator of $H_B(r)$ if Ξ is ergodic and B is a star-shaped compact set [Hei93, Lemma 3]. The estimator $\hat{H}_{B,W}(r)$ is also asymptotically normal. The isotropic case is treated in the following theorem.

Theorem 4.8 (see [Hei93]) *Let Ξ be a stationary Boolean model in $\mathbf{R}^d$ with intensity λ and convex compact isotropic typical grain Ξ_0 satisfying $\mathbf{E}(V_k(\Xi_0))^2 < \infty$ for $k = 0, \ldots, d-1$. Further, assume that B is a convex subset of $\mathbf{R}^d$ containing the origin. Then the finite-dimensional distributions of the process*

$$\hat{\zeta}_W(r) = \mathrm{mes}(W)^{1/2}(\hat{H}_{B,W}(r) - H_B(r)), \quad 0 \leq r \leq r_0,$$

converge as $W \uparrow \mathbf{R}^d$ to the finite-dimensional distributions of a Gaussian centred random process $\zeta(r)$, $0 \leq r \leq r_0$, with the covariance function

$$\mathbf{E}[\zeta(s)\zeta(t)] = \frac{1}{(1-p)^2} \int_{\mathbf{R}^d} \Big[\exp\{-\lambda\mathbf{E}\mathrm{mes}((\Xi_0 \oplus s\check{B}) \cup (\Xi_0 \oplus t\check{B} + x))\}$$

$$- \exp\{-\lambda\mathbf{E}\mathrm{mes}((\Xi_0 \oplus s\check{B}) \cup (\Xi_0 + x))\}$$

$$- \exp\{-\lambda\mathbf{E}\mathrm{mes}((\Xi_0 \oplus t\check{B}) \cup (\Xi_0 + x))\}$$

$$+ \exp\{-\lambda\mathbf{E}\mathrm{mes}(\Xi_0 \cup (\Xi_0 + x))\}\Big] dx.$$

The additional condition

$$\sup_{\substack{0 \leq s \leq t \leq r_0 \\ t-s \leq h}} \mathbf{E}\Big[\mathrm{mes}(\Xi_0 \oplus t\check{B})^3 \mathrm{mes}\big((\Xi_0 \oplus t\check{B}) \setminus (\Xi_0 \oplus s\check{B})\big)\Big] \leq c(r_0, B)h \qquad (4.21)$$

for $h > 0$ with a constant $c(r_0, B)$ depending on r_0 and B only, yields the functional convergence of $\hat{\zeta}_W(\cdot)$ in the Skorokhod space $\mathcal{D}[0, r_0]$ (see [Bil68]) to the a.s. continuous Gaussian process $\zeta(\cdot)$ with the covariance given above.

For the spherical contact distribution function, (4.21) follows from

$$\mathbf{E}[\mathrm{mes}(\Xi_0^r)^3 V_j(\Xi_0^r)] < \infty$$

for some $r > 0$.

Kaplan–Meier estimators. The spherical contact distribution function is closely related to the so-called *empty space function* given by

$$F(r) = \mathbf{P}\{\rho(o, \Xi) \leq r\},$$

where the origin o can be replaced by an arbitrary fixed point $x \in \mathbf{R}^d$. This function is often used in statistics of point processes [SKM95]. Clearly,

$$H_o(r) = F(r)/(1 - F(0)) \, .$$

The reduced-sample estimator is based on (4.1) and uses the observation of Ξ inside the window $W \ominus B_r(o)$. The so-called *Kaplan–Meier estimator* suggested by Baddeley and Gill [BG93] is defined by

$$\hat{F}^{\mathrm{km}}(r) = 1 - \exp\left\{ -\int_0^r \frac{\mu_{d-1}(\partial(\Xi \oplus sB) \cap (W \ominus sB))}{\mu_d((W \ominus sB) \setminus (\Xi \oplus sB))} ds \right\} \, , \tag{4.22}$$

where B is the unit ball and μ_{d-1} is the $(d-1)$-dimensional Hausdorff measure (surface area). The interpretation of the set $(W \ominus sB) \setminus (\Xi \oplus sB)$ as 'points at risk', and $\partial(\Xi \oplus sB) \cap (W \ominus sB)$ as observed 'deaths' establishes links to the classical Kaplan–Meier estimator from survival analysis [GJ90]. Further estimators are discussed in [CS97].

The estimator $\hat{F}^{\mathrm{km}}(r)$ is a monotone function. It should be noted that in most natural examples the asymptotic properties of the Kaplan–Meier estimator and the minus-sampling estimator of F do not differ very much (see [BG94]). However, artificial examples show that the Kaplan–Meier estimator can be considerably better.

Similar construction works also for general stationary regular closed random set Ξ [BG94] and for linear contact distribution functions [Han95, HGB96].

Morphological statistics. A general approach to defining statistics based on morphological operations was suggested in [Rip86] and elaborated in [MA91]. Reproduced below (in our notation) are some results of [MA91].

Let X be a general stationary ergodic random closed set. Define two statistics, $S_W(r)$ and $U_W(r)$, depending on $r > 0$, by

$$S_W(r) = \frac{\mathrm{mes}((X \bmod K_r) \cap (W \circ K_r))}{\mathrm{mes}(W \circ K_r)} \tag{4.23}$$

and

$$U_W(r) = \frac{\mathrm{mes}((X \bmod K_r) \cap (W \circ K_r))}{\mathrm{mes}(X \cap (W \circ K_r))} \, , \tag{4.24}$$

where $\bmod K_r$ denotes a fixed morphological operation using a structuring element K_r of 'size' r. The 'minus-window' $W \circ K_r$ consists of points $x \in W$ for which it can be verified whether or not the random set $X \bmod K_r$ contains x. For example, if $X \bmod K_r = (X \ominus \check{K}_r) \oplus K_r$ is the *opening* by K_r [Mat75, Ser82], then $W \circ K_r = W \ominus (K_r \oplus \check{K}_r)$. If $X \bmod K_r = X \oplus K_r$ is the dilation by K_r, then $S_W(r)$ is the empirical capacity functional $\hat{T}_{X,W}(K_r)$.

It is easy to prove that $\mathbf{E}S_W(r) = \mathbf{P}\{o \in X \bmod K_r\} = F(r)$ and

$$\mathrm{Var}\, S_W(r) = \frac{1}{\mathrm{mes}(W \circ K_r)^2} \iint_{W \circ K_r} \Big[\mathbf{P}\{\{x, y\} \subset X \bmod K_r\}$$

$$- \mathbf{P}\{o \in X \bmod K_r\}^2 \Big] dx\, dy \, .$$

Lemma 3.1 yields the asymptotic variance as $W \uparrow \mathbf{R}^d$

$$\lim_{W \uparrow \mathbf{R}^d} \mathrm{mes}(W) \operatorname{Var} S_W(r) = \int_{\mathbf{R}^d} (\mathbf{P}\{\{o, v\} \subset X \bmod K_r\} - F(r)^2) dv \qquad (4.25)$$

provided the integral is finite. In particular, this gives asymptotic variance (4.3) of the empirical capacity functional. Furthermore, if the integral in (4.25) is finite, then S_W converges in probability to $F(r)$, and $U_W(r)$ converges in probability to $F(r)/F(0)$.

Further properties of such morphological statistics (convergence almost surely, the uniform convergence, the asymptotic normality) are studied in [MA91]. The uniformity with respect to the family of different structuring elements can be explored using the approach suggested in [Mol87], see also Theorem 4.5.

Generalised distances and edge correction. A general approach to defining contact distances was suggested by Rataj [Rat93]. Put

$$\sigma_B(x, X) = \sup\{r \geq 0 : (x + rB) \cap X \text{ is either } \{x\} \text{ or empty}\} . \qquad (4.26)$$

Given a distribution ξ on $\mathbf{R}^d \times \mathcal{F}$, the distribution of $\xi \sigma_B^{-1}$ of contact distances can be considered. In other words, both x and X in (4.26) can be random. In particular,

$$\xi(d(x, X)) = \mathbf{1}_{x \in X^c} \mathrm{mes}(dx) P(dX)$$

yields the contact distribution function with the structuring element B. Here P is the distribution of the corresponding random closed set, and $d(x, X)$ and dX are differentials in $\mathbf{R}^d \times \mathcal{F}$ and $\mathcal{F}$ respectively.

Furthermore,

$$\xi(d(x, X)) = \mathbf{1}_X \mathrm{N}(dx) Q(dX)$$

gives rise to the nearest neighbour distance distribution of Q [SKM95, Section 4.4], where Q is the distribution of a point process and $\mathrm{N}(dx)$ is the counting measure. The measure

$$\xi(d(x, X)) = \mathbf{1}_{x \in X^c} \mathrm{mes}(dx) P(dX)$$

with $B = [o, u]$ generates the distribution of the visibility extent in the direction $u \in \mathbf{S}^{d-1}$ (the radius-vector function of the star), see [SKM95, Section 3.2] and p. 51. Further examples can be found in [Rat93]. These contact distances belong to infinite-order characteristics of random sets, i.e. they cannot be found through their finite-order parameters.

The value $\sigma_B(x, X)$ inside the window W can be estimated through its 'censored' analogues only, for example, by those considered in [Rat93],

$$\sigma_B^{W\#}(x, X) = \sigma(x, X \cup \partial W),$$

$$\sigma_B^{W*}(x, X) = \begin{cases} \sigma_B(x, X) & \text{if } \sigma_B(x, X) \leq \sigma_B(x, \partial W), \\ +\infty & \text{otherwise}. \end{cases}$$

These contact distances generate the distribution functions denoted by $F^{W\#}$ and F^{W*} in contrast to the function $F_B(r)$ generated by the non-censored contact distance.

Theorem 4.9 (see [Rat93]) *Let B be a compact set in $\mathbf{R}^d$ containing the origin, W another compact set with $\mathrm{mes}(W) > 0$ and ξ a Borel measure on $\mathbf{R}^d \times \mathcal{F}$ such that $\xi(K \times \mathcal{F}) < \infty$ for each compact set K and $\xi(d(x, X)) = \xi(d(x - y, X - y))$, $y \in \mathbf{R}^d$. Then we have*

$$F^{W\#}(r) = 1 - R_B^W(r) + R_B^W(r)F_B(r),$$

$$F^{W*}(r) = \int\limits_{[0,r]} R_B^W(s)F_B(ds), \quad r \geq 0,$$

where $R_B^W(s) = \mathrm{mes}(W)^{-1}\mathrm{mes}(W \ominus s\breve{B})$, $s \geq 0$.

4.4 Pair-Correlation Functions

The pair-correlation function $g(v)$ of the point process reveals its second-order characteristics and can be used to describe interactions between points of the point process, see [SKM95, pp. 129, 130], [SS94, pp. 249–258] and [DVJ88]. When multiplied by the square intensity, $g(v)$ is equal to the infinitesimal probability of having two points of the point process: one located in the neighbourhood of the origin, the other in the neighbourhood of v (see also p. 37).

The pair-correlation function of the Poisson point process of germs is identically equal to 1. However, this point process is not observable. Perhaps the most important observable point process related to the Boolean model Ξ is the point process $N^+(u)$ of tangent points in the direction u, see Section 3.2. Its pair-correlation function $g_u(v)$ can be estimated by means of a kernel estimator, say, given by

$$\hat{g}_{u,W}(v) = \frac{1}{(\hat{N}_{A,W}^+(u))^2} \sum_{\substack{x,y \in N^+(u) \cap W \\ x \neq y}} \frac{k_\varepsilon(x - y - v)}{\mathsf{A}(W \cap (W + y - x))}. \qquad (4.27)$$

Here k_ε is a kernel function that depends on parameter ε (a kind of bandwidth); for example, in the plane

$$k_\varepsilon(v) = \begin{cases} \frac{2}{\pi\varepsilon^2}\left(1 - \frac{\|v\|^2}{\varepsilon^2}\right) & , \quad \|v\| \leq \varepsilon, \\ 0 & , \quad \text{otherwise}. \end{cases} \qquad (4.28)$$

Estimator (4.27) is a modification of classical estimators of the pair-correlation functions used in the isotropic case, see [SKM95, p. 136] and [SS94, pp. 284–286]. It is not possible to apply estimators designed for isotropic point processes directly, since the point process of tangent points is *always* anisotropic [MS94a]. Unfortunately, asymptotic properties of kernel estimators for anisotropic point processes have not been systematically investigated.

The pair-correlation function $g_{u_1,u_2}(v)$ between tangent points with respect to two directions u_1 and u_2 can be estimated similarly to (4.27):

$$\hat{g}_{u_1,u_2,W}(v) = \frac{1}{\hat{N}^+_{A,W}(u_1)\hat{N}^+_{A,W}(u_2)} \sum_{\substack{x \in N^+(u_1) \cap W \\ y \in N^+(u_2) \cap W}} \frac{k_\varepsilon(x - y - v)}{A(W \cap (W + y - x))}.$$

(4.29)

Note that $\lambda^2(1-p)^2 g_{u_1,u_2}(v)$ is equal to the infinitesimal probability to have two tangent points in directions u_1 and u_2, one in the neighbourhood of the origin and the other one in the neighbourhood of v.

Another pair-correlation function is the pair-correlation function $g_{\partial\Xi}(v)$ of the fibre process $\partial\Xi$. It is of use when finding the asymptotic variance of the estimator of the specific boundary length, see (3.30). Its estimator for the isotropic fibre process has been considered in [SKM95, p. 296], [Sto81] and in the general case in [Han85, Sto85]. Fortunately, the fibre process $\partial\Xi$ is isotropic for each isotropic Boolean model Ξ.

Notes to Section 4.4

Estimation of pair-correlation functions in the isotropic case. Let N be an *isotropic* point process in $\mathbf{R}^d$ of intensity λ. We assume that the window of observations is fixed. Then the pair-correlation function $g(r)$ of N, see (p. 37) is estimated by

$$\hat{g}(r) = \frac{1}{2\pi r \lambda^2} \sum_{x,y \in N,\ x \neq y} \frac{k_\varepsilon(r - \|x - y\|)}{\mathrm{mes}((W + x) \cap (W + y))}.$$

For $d = 2$, the kernel function can be chosen as

$$k_\varepsilon(t) = \begin{cases} \frac{3}{4\varepsilon}\left(1 - \frac{t^2}{\varepsilon^2}\right) & , -\varepsilon \leq t \leq \varepsilon, \\ 0 & , \text{otherwise}. \end{cases}$$

Simulation experiments advise the use of the bandwidth

$$\varepsilon = c\lambda^{-1/2} \quad \text{with} \quad 0.1 \leq c \leq 0.2,$$

see [SS94, p. 285]. The isotropic assumption is used to construct the following estimator suggested by Ripley [Rip76] (see also [Ohs83]):

$$\hat{g}_{\mathrm{R}}(v) = \frac{1}{2\pi r \lambda^2} \sum_{x,y \in N,\ x \neq y} \frac{k_\varepsilon(r - \|x - y\|)b_{xy}}{\mathrm{mes}(W^{\|x-y\|})},$$

(4.30)

where $b_{xy} = 2\pi/\alpha_{xy}$, α_{xy} is the sum of all angles of the arcs in W of a circle with centre x and radius $\|x - y\|$, and $W^r = W \oplus B_r(o)$. Another estimator suggested by Ohser [Ohs83, OS81] is given by

$$\hat{g}_O(v) = \frac{1}{2\pi r \lambda^2 \bar{\gamma}_W(r)} \sum_{x,y \in N,\, x \neq y} k_\varepsilon(r - \|x - y\|),$$

where $\bar{\gamma}_W(r) = \mathbf{E}\mathrm{mes}(\omega W \cap (\omega W + r))$ is the isotropised set-covariance function of W defined for isotropic rotation ω. There is as yet no well-developed asymptotic theory for all these estimators. Some results are known for special types of point processes, for instance, for Poisson cluster processes [Hei88a].

Estimation of moment measures. Several results for estimators of moment measures and product densities of point processes are outlined below. The presentation follows the survey in [Jol91].

Let N be a stationary point process in $\mathbf{R}^d$. The kth-order factorial moment measure $\alpha^{(k)}(\cdot)$ (3.15) can be used to define the integral

$$\mu^{(k)}(g) = \mathbf{E} \int_{(\mathbf{R}^d)^k} g(x_1, \ldots, x_k) \alpha^{(k)}(d(x_1, \ldots, x_k))$$

for any real bounded measurable function g on $(\mathbf{R}^d)^k$, with compact range (cf. p. 37). The reduced kth-order moment measure $\alpha'^{(k)}$ is defined via the disintegration

$$\mu^{(k)}(g) = \int_{(\mathbf{R}^d)^k} g(u_1, \ldots, u_{k-1}, x_k) \alpha'^{(k)}(du_1 \cdots du_{k-1}) dx_k$$

with $u_i = x_i - x_k$ for $i = 1, \ldots, k-1$. Take $g(x_1, \ldots, x_k) = \mathbf{1}_{x_k \in W} h(x_1, \ldots, x_{k-1})$. A natural estimator of $\alpha'^{(k)}(h)$ is

$$\hat{\alpha}_W'^{(k)}(h, N) = \frac{1}{\mathrm{mes}(W)} \sum_{x_k \in N \cap W} \eta(x_k, N), \qquad (4.31)$$

where

$$\eta(y, N) = \int_{(\mathbf{R}^d)^{k-1}} h(x_1 - y, \ldots, x_{k-1} - y) N(dx_1) \cdots N(dx_{k-1}).$$

We recall that N denotes at the same time the realisation of a set of points and the corresponding counting measure. If $\eta(y, N)$ is an integrable random variable, then estimator (4.31) is unbiased, see [Kri82]. It follows from general results by Nguyen and Zessin [NZ76] that the integrability of $\eta(y, N)$ yields the strong consistency of the estimator (4.31) as $W \uparrow \mathbf{R}^d$.

The asymptotic normality of $\mathrm{mes}(W)^{1/2}(\hat{\alpha}_W'^{(k)}(h, N) - \mu'^{(k)}(h))$ can be proved under the so-called *Brillinger mixing* condition [Bri75, Jol80, Jol84] or for the so-called Poisson cluster processes [Hei88b]. The limiting Gaussian distribution has the variance determined by h and the $2p$ first moments of N. Moreover, the Brillinger mixing implies the functional convergence for the family of functions $h_t(\cdot)$ depending on a parameter t, see [Jol80]. The latter provides the possibility of investigating also kernel estimators obtained for

$$h_t(u) = \beta_t^{-d} k(\beta_t^{-1} u),$$

where $t \mapsto \beta_t$ is a non-negative function on the set of positive reals such that $\beta_t \to 0$ as $t \to \infty$. Further results including the speed of the convergence can be found in [Jol86, Jol91].

Estimation of the pair-correlation function of the fibre process. Note that, in contrast to the point process of tangent points, the fibre process $\partial\Xi$ is isotropic for an isotropic Boolean model Ξ. We will consider a general isotropic fibre process Φ. Its second-order properties are described by the second reduced moment function $K(r)$, where $L_A K(r)$ is the mean length of all fibre pieces within the disk (ball) $B_r(x)$, L_A is the intensity of the fibre process and x is a 'typical' fibre point, see also p. 41. Other second-order characteristics of fibre processes are discussed in [MS80, Sto81, SKM95, SMP80]. If the second reduced moment function K has a derivative, then

$$g(r) = \frac{K'(r)}{d b_d r^{d-1}}, \quad r \geq 0,$$

is the pair-correlation function of the fibre process. This function can be used for qualitative analysis of fibre patterns. For example, the speed of convergence $g(r)$ to 1 as $r \to \infty$ characterises the 'randomness' of the fibre process.

Below we enlist several estimation methods for the functions K and g. The most accurate direct estimator of $K(r)$ is given by

$$\hat{K}_W(r) = \left[\mu_1((W \ominus B_r(o)) \cap \Phi)\right]^{-1} \int\limits_{(W \ominus B_r(o)) \cap \Phi} \mu_1(B_r(x) \cap \Phi)\mu_1(dx),$$

where the integration is taken with respect to the 1-dimensional Hausdorff measure μ_1 (along the fibres). This method can be simplified by taking a test system of lines and calculating

$$\hat{K}_W(r) = n^{-1} \sum_{i=1}^{n} \mu_1(B_r(x_i) \cap \Phi \cap W)$$

for the points $x_1, \ldots, x_n$ resulting from the intersections of Φ and the test system of lines within the minus-window $W \ominus B_r(o)$.

Hanisch [Han85] suggested the so-called line intercept method to estimate $g(r)$. For this, the fibres must be approximated by segments. On each segment, independently of others, we must generate a Poisson point process of the (linear) intensity 1. The union of all such points can be interpreted as a Cox process (doubly stochastic Poisson process [DVJ88, SKM95]) with driving measure $\Phi(\cdot)$ such that $\Phi(B)$ is equal to the length of the fibre system inside B. Then the pair-correlation function of this Cox process and the fibre process coincide, so that an estimate of the first pair-correlation function can be used to estimate the second one.

Another approach uses intersection points of the fibre process with a Poisson line process Φ_0 of intensity λ. The latter is defined as on p. 12 for $d = 2$. The intersection $\Phi \cap \Phi_0$ is a stationary point process of intensity $2\lambda L_A/\pi$. Its pair-correlation function is related to the pair-correlation function g of Φ, see [SKM95, p. 296]. Then $(2\lambda L_A/\pi)^2 K(r)$ is estimated by

$$\sum_{\substack{x,y \in \Phi \cap \Phi_0 \cap W \\ 0 < \|x-y\| \le r}} \frac{k(x,y)}{\mathsf{A}(W^{\|x-y\|})},$$

where $k(x,y) = 2\pi/\alpha_{xy}$, and α_{xy} is the sum of all angles of the arcs in W of a circle with centre x and radius $\|x - y\|$, see [SKM95, p. 135]. The summation here is performed for pairs of points arising from different lines of the Poisson line process Φ_0.

Unfortunately, apart from strong consistency and unbiasedness, nothing is known about asymptotic properties of the enlisted estimators. The case of non-isotropic fibre processes is discussed in [Sto85].

5

Estimation of Numerical Individual Parameters

The intensity λ is the most important individual parameter of the Boolean model. In fact, statistics of the Boolean model began with intensity estimation. Further numerical parameters include mean values of other numerical functionals of the grain, for example, the mean area $\bar{A} = \mathbf{EA}(\Xi_0)$ and the mean perimeter $\bar{U} = \mathbf{EU}(\Xi_0)$ in the planar case.

The section headings of this chapter provide a kind of classification of existing estimation methods. Similar classifications and reviews can be found in [LS91, Mol95, MS94a, MS94b, Sch92].

5.1 Minimum Contrast Method

The *minimum contrast method* for contact distribution functions which was introduced in Section 4.3 dates back to [Dig81, Dup80] and [Ser82]. It is sometimes called the Steiner method [PS88, Sch92].

Let the structuring element B be the unit disk, i.e. consider the spherical contact distribution function $H_\circ(r)$. Then the corresponding logarithmic transform, i.e. the logarithmic spherical contact distribution function, is equal to

$$
\begin{aligned}
H_\circ^l(r) &= -\log(1 - H_\circ(r)) \\
&= \lambda \Big[\mathbf{EA}(\Xi_0 \oplus rB) - \mathbf{EA}(\Xi_0) \Big].
\end{aligned}
$$

The Steiner formula (2.26) from the integral geometry [Mat75, p. 81], [SKM95, pp. 13, 199] expresses the measure of the dilated set $\Xi_0^r = \Xi_0 \oplus rB$ through the values of the geometric functionals of Ξ_0. In the planar case for the convex set Ξ_0 it is

$$A(\Xi_0^r) = \pi r^2 + rU(\Xi_0) + A(\Xi_0) \,,$$

whence

$$H_\circ^l(r) = \lambda \pi r^2 + \lambda r \bar{U} \,.$$

Therefore, we can conclude that the function

$$\frac{H_\circ^l(r)}{r} = \lambda \pi r + \lambda \bar{U} \tag{5.1}$$

is linear. The essence of the minimum contrast method lies in the finding of a linear function $f(r) = ar + b$, which approximates the empirical logarithmic spherical contact distribution function

$$r^{-1} \hat{H}_{\circ,W}^l(r) = -r^{-1} \log(1 - \hat{H}_{\circ,W}(r)) \,.$$

We say that the contrast between $f(r)$ and $r^{-1}\hat{H}_{\circ,W}^l(r)$ is minimum. For the measure of contrast it is possible to use the uniform distance,

$$\sup_{0 \le r \le r_0} |f(r) - r^{-1}\hat{H}_{\circ,W}^l(r)|$$

for $r_0 > 0$, or some integral metrics, for example,

$$\int_0^{r_0} (f(r) - r^{-1}\hat{H}_{\circ,W}^l(r))^2 \, dr \,, \tag{5.2}$$

see [Cre91, Hei93]. It is possible also to minimise the contrast between the empirical spherical contact distribution function $\hat{H}_{\circ,W}(r)$ and the exponential function $\exp\{-(\lambda \pi r^2 + \lambda \bar{U} r)\}$ rather than the contrast between their logarithms. It should be noted that this is not a classical regression problem, since the values of $\hat{H}_{\circ,W}(r)$ for different r are not independent.

If the 'best' linear function (linear approximator of $r^{-1}\hat{H}_{\circ,W}^l(r)$)

$$\hat{f}_W(r) = \hat{a}_W r + \hat{b}_W$$

has been found, then the estimators of λ and $\bar{U}$ can be obtained from the representation (5.1) as

$$\hat{\lambda}_W = \hat{a}_W/\pi \,, \quad \hat{\bar{U}}_W = \hat{b}_W \pi/\hat{a}_W \,.$$

It follows from (2.12) that the mean area can be estimated using $\hat{\lambda}_W$ and $\hat{p}_W$ as

$$\hat{\bar{A}}_W = -\hat{\lambda}_W^{-1} \log(1 - \hat{p}_W) \,.$$

The minimum contrast estimators with respect to the integral contrast metric (5.2) are investigated in [Hei93]. These estimators are strong consistent and asymptotically normal. Unfortunately, their asymptotic variances are very complicated even for simple Boolean models.

In higher dimensions the order of the polynomial in (5.1) gets larger, but it is still possible to find estimators of the geometric functionals of the grain [Cre91, Hei93, Sch92].

On the discretised screen the evaluation of the spherical contact distribution function may cause problems for small r, since the quality of approximation of small smooth circles is very low for a discrete grid. Then it is advisable to use the quadratic contact distribution function. If the grain is *isotropic* and the structuring element B is the unit square, then we get instead of (5.1)

$$\frac{H_\square^l(r)}{r} = \lambda r + \frac{2}{\pi}\lambda\bar{U}\,.$$

For a general contact distribution function the isotropy assumption yields

$$\frac{H_B^l(r)}{r} = \lambda\mathsf{A}(B)r + \lambda\frac{\mathsf{U}(B)}{2\pi}\bar{U}\,. \tag{5.3}$$

For example, it is advisable to take a triangle as a structuring element if a triangular sampling grid is used.

A variant of the minimum contrast method for the linear contact distribution function is discussed in Section 6.1. The application of the minimum contrast method to the covariance is also possible, but it is not easy because of the complicated theoretical form of the covariance function.

The evaluations of contact distribution functions involve area measurements of the set $\Xi\oplus rB$. Another approach can be based on the measurements of the specific boundary length of $\Xi\oplus rB$. For example, if B is the unit ball, then, by (2.13), the specific boundary length of $\Xi\oplus rB$ is given by

$$L_A(r) = \lambda(1 - p(r))(2\pi r + \bar{U})\,,$$

where $p(r)$ is the area fraction of the set $\Xi\oplus rB$. This makes it possible to estimate the intensity and the mean perimeter of the typical grain using the minimum contrast approach for the function $L_A(r)$, $r > 0$.

Notes to Section 5.1

Minimum contrast approach for capacity functionals. It is possible to consider a general minimum contrast approach for the capacity functional of the Boolean model [Mol92]. Let the capacity functional $T_\Xi(K, \theta)$ of Ξ depend on a parameter $\theta \in \Theta$. Then the minimum contrast estimator of θ is given by

$$\hat{\theta}_W = \arg\inf_{\theta\in\Theta}\ \sup_{K\in\mathfrak{M},\,K\subseteq K_0} |T_\Xi(K, \theta) - \hat{T}_{\Xi,W}(K)|$$

for a certain compact K_0 and a class of appropriate compact sets $\mathfrak{M}$ (e.g., the family of all disks or all squares). In particular cases this yields the minimum contrast method for the contact distribution functions.

Minimum contrast method in $\mathbf{R}^3$. It follows from the general Steiner formula (2.26) that in three-dimensional space

$$\mathsf{mes}(\Xi_0^r) = \frac{4}{3}\pi r^3 + \pi r^2 V_1(\Xi_0) + r\mathsf{S}(\Xi_0) + \mathsf{mes}(\Xi_0),$$

where $\mathsf{S}(\Xi_0)$ is the surface area of Ξ_0 and

$$V_1(\Xi_0) = \frac{1}{2\pi} \int\limits_{\mathbf{S}^2} h(\Xi_0, u)du = \bar{b}(\Xi_0),$$

where $\bar{b}(\Xi_0)$ is the mean width of Ξ_0 (the width $b(\Xi_0, u) = h(\Xi_0, u) + h(\Xi_0, -u)$ averaged for all directions $u \in \mathbf{S}^2$). Then

$$\frac{H_0^l(r)}{r} = \frac{4}{3}\pi r^2 \lambda + \pi r \lambda \mathbf{E}\bar{b}(\Xi_0) + \lambda \mathbf{E}\mathsf{S}(\Xi_0).$$

This allows us to estimate parameters by means of quadratic approximations of the empirical logarithmic spherical contact distribution function.

Asymptotic theory for the minimum contrast method. Suppose that the distribution of Ξ_0 depends on some parameter $\theta = (\theta_0, \ldots, \theta_m)$ that belongs to a compact set $\Theta \subset \mathbf{R}^{m+1}$. For example, θ can be a vector comprised of λ, $\mathbf{E}U(\Xi_0)$ and $\mathbf{E}A(\Xi_0)$.

Then the contact distribution function of Ξ with structuring element B depends on θ and is denoted by $H_B(r; \theta)$. The corresponding area fraction is denoted by $p(\theta)$. Heinrich [Hei93] used the contrast function

$$U_W(\theta) = \int\limits_0^{r_0} \left(\log \frac{1 - H_B(r; \theta)}{1 - \hat{H}_{B,W}(r)} \right)^2 G(dr) + g_0 \left(\log \frac{1 - p(\theta)}{1 - \hat{p}_W} \right)^2,$$

where g_0 and r_0 are some positive constants, and $G(\cdot)$ is a finite measure on $[0, r_0]$ with $G(\{0\}) = G(\{r_0\}) = 0$.

Let B be a convex structuring element such that $o \in B$ and

$$l = \max\{k \geq 0 : V_{d-k}(B) > 0\} \geq 1,$$

where $V_j(B)$ are intrinsic volumes, see p. 26. Use the notation $v_j(\theta) = \mathbf{E}_\theta V_j(\Xi_0)$ with the expectation taken with respect to the distribution corresponding to $T_\Xi(\cdot; \theta)$. Assume that the functions $v_j(\theta)$, $j = 1, \ldots, d$, are continuous on Θ and $v_j(\theta) = v_j(\theta')$, $1 \leq j \leq d$, implies $\theta = \theta'$. Finally, suppose that there exist $d - l$ distinct points $r_1, \ldots, r_{d-l} \in (0, r_0)$ such that

$$\min_{1 \leq j \leq d-l} G((r_j - \varepsilon, r_j + \varepsilon)) > 0$$

for any $\varepsilon > 0$. Then the minimum contrast estimator

$$\hat{\theta}_n = \arg \inf_{\theta \in \Theta} U_{W_n}(\theta)$$

converges to θ almost surely as $W_n \uparrow \mathbf{R}^d$, see [Hei93]. To formulate a limit theorem assume that the functions $v_j(\theta)$ are twice continuously differentiable on Θ, and $\mathbf{E}_\theta V_j(\Xi_0)^2 < \infty$ for all $\theta \in \Theta$ and $1 \leq j \leq d$. Define

$$A_i(r, \theta) = \frac{\partial}{\partial \theta_i} \log(1 - H_B(r, \theta)),$$

$$B_i(\theta) = g_0 \frac{\partial}{\partial \theta_i} \log(1 - p(\theta)) - \int_0^{r_0} \frac{\partial}{\partial \theta_i} \log(1 - H_B(r, \theta)) G(dr),$$

$$C_B(s, t, \theta) = \int_{\mathbf{R}^d} \left[\exp\{\lambda \mathbf{Emes}((\Xi_0 \oplus s\check{B}) \cap (\Xi_0 \oplus (s\check{B} + x)))\} - 1 \right] dx,$$

and assume that the matrix $s(\theta) = (s_{ij}(\theta))_{ij=0}^m$ with

$$s_{ij}(\theta) = \lim_{W_n \uparrow \mathbf{R}^d} \frac{\partial^2}{\partial \theta_i \partial \theta_j} \frac{U_{W_n}(\theta)}{2} \quad \text{in probability}$$

$$= \int_0^{r_0} A_i(r, \theta) A_j(r, \theta) G(dr) + g_0 \frac{\partial}{\partial \theta_i} \log(1 - p(\theta)) \frac{\partial}{\partial \theta_j} \log(1 - p(\theta))$$

is non-singular for all $\theta \in \Theta$.

Then the minimum contrast estimator $\hat{\theta}_n$ is asymptotically normal: namely, $\text{mes}(W_n)^{1/2}(\hat{\theta}_n - \theta)$ converges in distribution to a centred Gaussian random vector with the covariance matrix $(\sigma_{ij}(\theta))_{ij=0}^m$ given by

$$\sigma_{ij}(\theta) = \lim_{n \to \infty} \text{mes}(W_n) \mathbf{E} \left[\frac{\partial}{\partial \theta_i} \frac{U_{W_n}(\theta)}{2} \frac{\partial}{\partial \theta_j} \frac{U_{W_n}(\theta)}{2} \right]$$

$$= \int_0^{r_0} \int_0^{r_0} C_B(s, t, \theta) A_i(s, \theta) A_j(t, \theta) G(ds) G(dt) + C_B(0, 0, \theta) B_i(\theta) B_j(\theta)$$

$$+ \int_0^{r_0} C_B(0, r, \theta) [A_i(r, \theta) B_j(\theta) A_j(r, \theta) B_i(\theta)] G(dr),$$

see [Hei93].

5.2 Method of Intensities

This method is similar to the method of moments in classical statistics: namely, the estimators of parameters for the Boolean model are chosen to match the empirical values of aggregate parameters (or intensities of some functionals). When applied to anisotropic Boolean models, it is sometimes called Weil's method.

One starts with a functional of interest ϕ, such that its values on convex sets are well-defined. This functional is extended additively onto the convex ring (family of finite unions of convex sets). This additive extension means that

$$\phi(K_1 \cup K_2) = \phi(K_1) + \phi(K_2) - \phi(K_1 \cap K_2)$$

for all convex compact sets K_1 and K_2. Similarly, $\phi(\Xi \cap W)$ can be defined, since $\Xi \cap W$ is a union of a finite number of convex sets. Then the *spatial intensity* (or spatial density)

$$D_\phi = \lim_{W \uparrow \mathbf{R}^2} \frac{\phi(\Xi \cap W)}{\mathrm{mes}(W)}$$

is determined. The principal point is to find an expression of D_ϕ through individual parameters of the Boolean model.

In the plane it is usual to take two standard functionals and the corresponding equations:

$$p \;=\; 1 - \exp\{-\lambda\bar{A}\}, \tag{5.4}$$
$$L_A \;=\; \lambda(1-p)\bar{U}. \tag{5.5}$$

They are obtained with ϕ being the area and the boundary length respectively. On the left-hand sides appear the area fraction, p, and the specific boundary length, L_A.

These equations together with formula (2.14) for the specific connectivity number (Euler–Poincaré characteristic) make it possible to express λ, $\bar{U}$ and $\bar{A}$ through p, L_A and χ_A. Remember that if the grain is isotropic, then

$$\chi_A = (1-p)\left(\lambda - \frac{\lambda^2}{4\pi}\bar{U}^2\right). \tag{5.6}$$

Then (5.5) and (5.6) can be solved to get

$$\lambda = \frac{\chi_A}{1-p} + \frac{1}{4\pi}\frac{L_A^2}{(1-p)^2} \tag{5.7}$$

and

$$\bar{U} = \frac{4\pi L_A(1-p)}{4\pi(1-p)\chi_A + L_A^2}. \tag{5.8}$$

The last step is to replace the latter aggregate parameters by their empirical counterparts discussed in Chapter 3.

Similarly to the method of moments in classical statistics the method of intensities leads to biased, but strong consistent results. Also empirical studies [LS91, Sch92] show its good efficiency. However, the exact asymptotic properties of the corresponding estimators are not known. While the asymptotic normality can be deduced, it is not easy to express the variance of the limiting distribution through parameters of the grain. This is explained by the lack of second-order integral-geometric formulae, see [Wei83] and p. 48.

Formula (5.6) is valid only in the isotropic case. In general it should be replaced by

$$\chi_A = (1 - p)\left(\lambda - \lambda^2 A(E\Xi_0, -E\Xi_0)\right), \qquad (5.9)$$

see [Wei88, Wei95]. Here $A(\cdot, \cdot)$ is the so-called *mixed area*, see (3.40) and [Sch93b, SW92], $E\Xi_0$ is the Aumann expectation of the grain, and $-E\Xi_0 = \{-x : x \in E\Xi_0\}$. Note that (5.9) is valid also for non-convex grains and yields an estimator for the intensity in a general framework, see Section 6.2.

Notes to Section 5.2

Method of intensities in $\mathbf{R}^3$. Three-dimensional variants of the method can be found in [Sch92, Wei88, Wei95]. If Ξ_0 is isotropic, then one can use the system of equations

$$
\begin{aligned}
p &= 1 - \exp\{-\lambda\bar{V}\}, \\
S_V &= (1 - p)\lambda\bar{S}, \\
M_V &= (1 - p)(\lambda\bar{M} - \pi^2\lambda^2\bar{S}^2/32), \\
\chi_V &= (1 - p)\left(\lambda - \frac{1}{4\pi}\lambda^2\bar{M}\bar{S} + \frac{\pi}{384}\lambda^3\bar{S}^3\right),
\end{aligned}
$$

with the left-hand sides obtained as spatial densities of the volume, surface area, integral of the mean curvature and the Euler–Poincaré characteristic. Furthermore, $\bar{S}$ is the expected surface area of the grain and $\bar{M}$ is the expected integral of the mean curvature of Ξ_0, or, equivalently, $\bar{M}$ is the expected mean width of Ξ_0 divided by the volume of the unit ball in $\mathbf{R}^3$, $b_3 = 4\pi/3$.

Without the isotropy assumption the method of intensities is much more difficult to use, see [Sch92, Wei88].

5.3 Tangent Points Method

This method uses the same two equations (5.4) and (5.5) as the method of intensities. However, the third equation deals with the specific *convexity* number instead of the specific *connectivity* number. Three equations (5.4), (5.5) and (3.10) make it possible to estimate the three main numerical parameters in the planar case. The estimator of the intensity is especially simple:

$$\hat{\lambda}_W = \frac{\hat{N}_{A,W}^+}{1 - \hat{p}_W},$$

(5.10)

where $\hat{N}_{A,W}^+$ is the empirical intensity of the point process of tangent points (with respect to an arbitrary direction u) considered in Section 3.2 and $\hat{p}_W$ is the empirical area fraction, see Section 3.1. Note that this estimator for the intensity keeps its form for any dimension and does not require the isotropy assumption. Estimator (5.10) together with (5.4) and (5.5) yields estimators of the mean perimeter and the mean area of the grain.

Estimator (5.10) is strong consistent and has very simple asymptotic properties. While the asymptotic variances of $\hat{p}_W$ and $\hat{N}_{A,W}^+$ (see (3.2) and (3.12)) are rather complicated, the limit theorem for $\hat{\lambda}_W$ is very simple.

Assume that the diameter of the grain and its area have finite second moments (in $\mathbf{R}^d$ the dth moment of the diameter and the second moment of the volume of the grain must be finite). Then the random variable (normalised difference)

$$A(W)^{1/2}(\hat{\lambda}_W - \lambda)$$

converges in distribution to a Gaussian random variable with zero mean and the variance

$$V_\lambda = \frac{\lambda}{1 - p},$$

(5.11)

see [MS94a]. From this, it is possible to build confidence intervals for the intensity. The estimator $\hat{\lambda}_W$ and its asymptotic properties do not depend on the shape of the grain, the isotropy assumption, the direction u used to define tangent points, or the dimension of the space.

The estimator $\hat{\lambda}_W = \hat{\lambda}_W(u)$ can be improved by averaging with respect to different directions u used to define tangent points. For example, the estimator

$$(2\pi)^{-1} \int\limits_0^{2\pi} \hat{\lambda}_W(u) du$$

(5.12)

can be used. Further results can be found in [Mol95], see also p. 85.

Notes to Section 5.3

Asymptotic variance. The asymptotic normality can be deduced using the truncation approach described on p. 32. To derive the variance (5.11), we note that

$$\hat{\lambda}_W - \lambda = \frac{(\hat{N}_{A,W}^+(u) - N_A^+)(1 - p) - (\hat{p}_W - p)N_A^+}{(1 - \hat{p}_W)(1 - p)},$$

(5.13)

where $\hat{N}^+_{A,W}(u)$ is the number of tangent points in direction u divided by the volume of the window. The denominator tends almost surely to $(1-p)^2$ as $W \uparrow \mathbf{R}^d$. Furthermore,

$$\lim_{W \uparrow \mathbf{R}^d} \text{mes}(W)\left(\mathbf{E}\left[\hat{N}^+_{A,W}(u)\hat{p}_W - N^+_A p\right]\right) = \lambda \int_{\mathbf{R}^d} \left((1-p)^2 - (1-2p+C(v))\right)\psi_u(v)dv,$$

see [MS94a], where $\psi_u(b)$ has been defined in (3.14). Now the computation of the variance of the limiting numerator in (5.13) yields (5.11). For this one uses formulae (3.2) and (3.12) for the asymptotic variances of $\hat{p}_W$ and $\hat{N}^+_{A,W}$.

Note that the variance V_λ given by (5.11) is *not* the limiting normalised variance of $\hat{\lambda}_W$. The limit theorem just says that $\hat{\lambda}_W$ is asymptotically normal with variance (5.11). On the other hand, it is not easy to find the limit of $\text{mes}(W)\text{Var }\hat{\lambda}_W$ or even to establish its finiteness.

Averaged estimators for different directions. Let ν be a probability measure on $\mathbf{S}^{d-1}$. Then the estimator

$$\hat{\lambda}_W(\nu) = \int_{\mathbf{S}^{d-1}} \hat{\lambda}_W(u)\nu(du) \tag{5.14}$$

has a lower (or at most equal) variance than $\hat{\lambda}_W(u)$, $u \in \mathbf{S}^{d-1}$, since $\hat{N}^+_{A,W}(u)$, $u \in \mathbf{S}^{d-1}$, is a stationary process on the sphere, see also [SOS87]. Note that $\hat{\lambda}_W(u)$ is obtained for ν concentrated at $\{u\}$. Evidently,

$$\hat{\lambda}_W(\nu) = \frac{\hat{N}^+_{A,W}(\nu)}{1-\hat{p}_W},$$

where

$$\hat{N}^+_{A,W}(\nu) = \int_{\mathbf{S}^{d-1}} \hat{N}^+_{A,W}(u)\nu(du).$$

The following result can be derived from Theorem 3.3 and uses the same notations.

Theorem 5.1 (see [Mol95]) *If $\mathbf{E}\|\Xi_0\|^d < \infty$ and $\mathbf{E}\text{mes}(\Xi_0)^2 < \infty$, then $\text{mes}(W)^{1/2}(\hat{\lambda}_W(\nu) - \lambda)$ converges in distribution to a centred Gaussian random variable with the variance*

$$V_\lambda(\nu) = \int_{\mathbf{S}^{d-1}} \int_{\mathbf{S}^{d-1}} \nu(du_1)\nu(du_2)$$

$$\times \left[\lambda^2 \int_{\mathbf{R}^d} q(v)[1 - \psi_{u_1}(v)][1 - \psi_{u_2}(-v)]dv + \lambda \mathbf{E}q(\xi_{u_1,u_2})\right].$$

Note that for $u_1 = u_2 = u$ the value $[1 - \psi_{u_1}(v)][1 - \psi_{u_2}(-v)]$ is identically zero, $\xi_{u,u} = o$, and $q(o) = 1/(1-p)$, whence (5.11) immediately follows from Theorem 5.1 for ν concentrated at $\{u\}$.

Consider now a particular case. Suppose that ν is concentrated (with weights $1/2$) at two symmetric points $\{u, -u\}$. Then $\xi_{u,-u}$ is the vector that joins the tangent points of Ξ_0 in the directions u and $-u$, so that

$$V_\lambda(\nu) = \frac{\lambda^2}{2} \int_{\mathbf{R}^d} q(v)[1 - \psi_u(v)]^2 dv + \frac{\lambda}{2} \mathbf{E}q(\xi_{u,-u}) + \frac{\lambda}{2(1-p)} \; .$$

For instance, let $\Xi_0 = B_r(o)$ be a deterministic ball centred at the origin. Then $\Xi_0^u = B_r(ru)$, $\xi_{u,-u} = 2ru$ almost surely, and $q(2ru) = 0$. Hence

$$V_\lambda(\nu) = \frac{\lambda}{2(1-p)} \left(1 + \lambda r^d \int_{B_1(u)} \exp\{-\lambda r^d (b_d - \gamma_{B_1}(w))\} dw \right) ,$$

where $\gamma_{B_1}(w) = \mathsf{mes}(B_1 \cap (B_1 + w))$. In particular, for $d = 1$ we get

$$V_\lambda(\nu) = \frac{\lambda}{1-p}(1 - \frac{1}{2}e^{-2\lambda r}) \le \frac{\lambda}{1-p} = V_\lambda \; .$$

Flat pieces on the boundary. Suppose that $d = 2$ and Ξ_0 may have a flat piece on its boundary orthogonal to u, i.e. the support set $\partial_u \Xi_0$ (see p. 36) may contain a segment. In this case the tangent point is defined to be one of the end-points of this segment, for example, one can take the lower-left tangent point. Another possibility is to count both end-points with weights, for example, attach to both points the same weight $1/2$. Then $\hat{N}_A^+$ is replaced by the average of intensities of exposed lower-left and lower-right tangent points. Let us denote the end-points of $\partial_u \Xi_0$ by $n'_u(\Xi_0)$ and $n''_u(\Xi_0)$. It follows from a general limit theorem in [HM95] that the corresponding estimator of the intensity is asymptotically normal with the variance

$$\frac{\lambda}{2} \left[\frac{1}{1-p} + \mathbf{E}q(n'_u(\Xi_0) - n''_u(\Xi_0)) \right] ,$$

which is less than $V_\lambda = 1/(1 - p)$, since $q(v) < q(o) = 1/(1 - p)$ for $v \ne o$.

5.4   Schmitt's Method

This method was suggested in [Sch91]. It is applicable to Boolean models with bounded grains, i.e. it is necessary to know that the grain Ξ_0 is contained almost surely within a disk of a fixed diameter r. On the other hand, the grain can be an arbitrary random compact set, there is no need to assume either convexity or connectivity of the grain.

Schmitt's estimator is based on the identity

$$\lambda = \frac{1}{\varepsilon^2} \log \frac{(1 - T_\Xi(G))(1 - T_\Xi(G \cup K \cup L))}{(1 - T_\Xi(G \cup K))(1 - T_\Xi(G \cup L))} , \tag{5.15}$$

which is valid for any $\varepsilon > 0$. For this, the sets G, K and L must be chosen in a special way, for example, the sets shown in Figure 5.1 will do. Note that these sets are quite large if the maximum size, r, of the grain is large. This causes larger variances of this estimator observed in simulation experiments, since the values of the capacity functional on the test sets are close to 1, so that the corresponding logarithms behave rather irregularly.

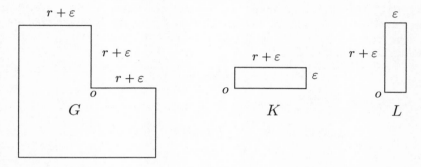

Figure 5.1   The sets used in the definition of Schmitt's estimator.

An estimator of λ can be easily obtained by replacing the capacity functional T_Ξ in (5.15) with its empirical counterpart $\hat{T}_{\Xi,W}$:

$$
\hat{\lambda}_W = \frac{1}{\varepsilon^2} \log \frac{(1 - \hat{T}_{\Xi,W}(G))(1 - \hat{T}_{\Xi,W}(G \cup K \cup L))}{(1 - \hat{T}_{\Xi,W}(G \cup K))(1 - \hat{T}_{\Xi,W}(G \cup L))} \tag{5.16}
$$

$$
= \varepsilon^{-2} \Big[\hat{\Psi}_{\Xi,W}(G \cup K) + \hat{\Psi}_{\Xi,W}(G \cup L) - \hat{\Psi}_{\Xi,W}(G) \\
- \hat{\Psi}_{\Xi,W}(G \cup K \cup L) \Big],
$$

see (4.5).

In principle, it is possible to use an analogue of (5.15) to estimate the capacity functional $\mathbf{P}\{\Xi_0 \cap G' \neq \emptyset\}$ for an open set G'.

Notes to Section 5.4

Proof of (5.15).   We will follow the scheme of the proof in [Sch91] and [Sch92]. First, note that the map

$$
c : K \mapsto c(K) = (x_1, \ldots, x_d)
$$

with $x_i = \min\{y_i : y = (y_1, \ldots, y_d) \in K\}$ is continuous with respect to the Hausdorff metric. The Hausdorff metric on $\mathcal{K}$ generates the σ-algebra σ_K. We introduce the measure Θ on σ_K by $\Theta(\mathcal{K}_K) = \lambda\mathbf{E}\mathrm{mes}(\Xi_0 \oplus \check{K})$, where $\mathcal{K}_K = \{F \in \mathcal{K} : F \cap K \neq \emptyset\}$. Then

$$\Theta(\mathcal{K}^G_{K,L}) = \Theta(\mathcal{K}_{G\cup K}) + \Theta(\mathcal{K}_{G\cup L}) - \Theta(\mathcal{K}_G) - \Theta(\mathcal{K}_{G\cup L\cup K}),$$

where

$$\mathcal{K}^G_{K,L} = \{F \in \mathcal{K} : F \cap G = \emptyset,\ F \cap K \neq \emptyset,\ F \cap L \neq \emptyset\}$$

Since $\Theta(\mathcal{K}_K) = \Theta(\mathcal{K}_{K+x})$ for all x, the measure Θ can be disintegrated and represented as the product of measure $Q(\cdot)$ on $\mathcal{K}$ and the Lebesgue measure in $\mathbf{R}^d$. The measure $Q(\cdot)$ is, in fact, the distribution of the typical grain.

Furthermore, note that

$$\Theta(c^{-1}(H)) = \lambda\mathrm{mes}(H)$$

for any Borel set $H \subset \mathbf{R}^d$. Indeed,

$$
\begin{aligned}
\theta(c^{-1}(H)) &= \lambda \int\limits_{\mathbf{R}^d} \int\limits_{\mathcal{K}} \mathbf{1}_H(c(K+x))Q(dK)dx \\
&= \lambda \int\limits_{\mathcal{K}} \int\limits_{\mathbf{R}^d} \mathbf{1}_H(c(K)+x)Q(dK)dx \\
&= \lambda \int\limits_{\mathcal{K}} \mathrm{mes}(H)Q(dK) = \lambda\mathrm{mes}(K).
\end{aligned}
$$

Thus,

$$\Theta(\mathcal{K}^G_{K,L}) = \Theta(c^{-1}([0,\varepsilon]^2)) = \lambda\varepsilon^2,$$

whence (5.15) easily follows.

Asymptotic variance. It is possible to find the asymptotic variance of the estimator (5.16) using the limit theorem for empirical capacities and the formulae (4.3), (4.7) and (4.12) for the empirical capacity functionals and the corresponding functional Ψ_Ξ. Then the normalised difference $A(W)^{1/2}(\hat{\lambda}_W - \lambda)$ converges in distribution to the sum of four correlated Gaussian random variables with zero mean and the covariance matrix

$$\left[\sigma_\Psi(A_i, A_j)\right]_{1 \leq i,j \leq 4},$$

where $A_1 = G$, $A_2 = G \cup K \cup L$, $A_3 = G \cup K$, $A_4 = G \cup L$.

Unfortunately, such a limit theorem is of little practical value because of complicated integrals in (4.3) and (4.12). Consequently, the best value of the parameter ε in (5.16) and the best choice of sets G, K and L are still unknown.

Empirical studies. Since the theoretical asymptotic variance of intensity estimators is known only for the tangent points method, other estimation methods can be compared only numerically by means of simulations. Lantuejoul and Schmitt [LS91] compared three methods: minimum contrast for spherical contact distribution functions (or the Steiner method), the method of intensities and Schmitt's method.

They considered the planar Boolean model of intensity 0.1, whose primary grain is a square with the sides of length 2 and parallel to the coordinate axes. One hundred independent simulations carried out in a square window W of side 50 were used to compare the chosen estimation methods. From the scatter plots presented in [LS91] it is possible to conclude that the method of intensities provides the best results of the three methods compared. Unfortunately, the empirical variances were not given in [LS91], nor was comparison with the tangent point method undertaken. Note that for the described simulation framework the area fraction is $p = 1 - e^{-0.4}$, whence (5.11) gives

$$V_\lambda = A(W)^{-1} \frac{0.1}{1-p} \approx 5.97 \cdot 10^{-5} \,.$$

Thus, the standard deviation of the tangent points estimator is equal to $\sigma_\lambda = 7.7 \cdot 10^{-3}$. In general, the variance given by the tangent points method can serve as a benchmark for comparisons of other estimators.

Further comparisons for mean values for different estimators were performed by Schröder [Sch92] (unfortunately, the corresponding variances and scatter plots were not presented). He found that the method of intensities and the tangent points method provide the most satisfactory results for different area fractions and Boolean models with polygonal typical grains.

5.5 Other Methods

Other estimation methods mostly rely on the isotropy assumption. We begin with two methods suggested in [AFM89]. Consider two contact distribution functions $H_{B_1}(r)$ and $H_{B_2}(r)$ and the corresponding contact distances (4.15) defined for two structuring elements B_1 and B_2. Then the random variables (transformed contact distances)

$$\eta_i = 2\lambda \left(\frac{U(B_i)}{2\pi} \bar{U} \zeta_{B_i} + A(B_i) \zeta_{B_i}^2 \right) , \quad i = 1, 2 \,,$$

have χ^2-distributions with two degrees of freedom [Aya88]. Thus, the expected values of η_1 and η_2 are 2. This yields the system of two equations for unknown λ and $EU(\Xi_0)$. The moments $E\zeta_{B_i}$ and $E\zeta_{B_i}^2$ can be estimated either through empirical contact distribution functions (Section 4.3) or by averaging the empirical contact distances (4.15) measured for different points outside Ξ.

The other method (sometimes called *the testing sets method*) suggested in [AFM89] uses the following idea. Let us take three compact sets K_1, K_2 and K_3 with $K_1 = \{o\}$, K_2 being a segment of length t (its direction is not important due to the isotropy assumption) and K_3 being a convex compact set of positive area containing the origin. Then the Steiner formula for isotropic random compact sets [Mat75, p. 85], [SKM95, p. 199] yields

$$\begin{cases} T_\Xi(K_1) &=& p = 1 - \exp\{-\lambda\bar{A}\}\,, \\ T_\Xi(K_2) &=& 1 - \exp\{-\lambda\left(\bar{A} + \bar{U}t/\pi\right)\}\,, \\ T_\Xi(K_3) &=& 1 - \exp\{-\lambda\left(\bar{A} + \bar{U}\mathsf{U}(K_3)/(2\pi) + \mathsf{A}(K_3)\right)\}\,. \end{cases}$$

Again, replacing T_Ξ by the empirical capacity functional (it can be evaluated through morphological transformations of Ξ, see [AFM89, Rip86] and Section 4.1) yields estimators of the parameters $\bar{A}$, $\bar{U}$ and λ.

A similar idea for K_i being a ball of radius r_i, $i = 1, 2, 3$, was used in [Mol90c] to estimate the moments of the radius of the spherical typical grain. Moreover, this choice of testing compact sets makes it possible to estimate the intensity for non-isotropic Boolean models. If $K_1 = \{o\}$, $K_2 = B_r$ and $K_3 = B_{cr}$ for some $c > 1$, then

$$\begin{cases} \Psi_\Xi(K_1) &=& -\log(1 - p) = \lambda\bar{A}\,, \\ \Psi_\Xi(K_2) &=& \lambda\pi r^2 + \lambda\bar{U}r + \lambda\bar{A}\,, \\ \Psi_\Xi(K_3) &=& \lambda\pi c^2 r^2 + \lambda\bar{U}cr + \lambda\bar{A}\,. \end{cases}$$

This system of equations allows us to find λ, $\bar{A}$ and $\bar{U}$ through observable values of the functional $\Psi_\Xi(K_i)$, $i = 1, 2, 3$, see (4.4) and (4.5). We will call this method *the testing balls method*.

Notes to Section 5.5

Testing compacts in higher dimensions. Suppose that the typical grain is the ball of radius ξ in $\mathbf{R}^d$. For general d one can use $(d + 1)$ testing compact sets $K = B_{r_i}$ for different radii r_i, $i = 0, \ldots, d$. For instance, put $r_i = i$, $i = 0, \ldots, d$. Similarly to the two-dimensional case one can write system of equations with $\Psi_\Xi(B_{r_i})$ in the left-hand sides. Let $\hat{\lambda}_W$ denote the corresponding intensity estimator obtained by replacing Ψ_Ξ with $\hat{\Psi}_{\Xi,W}$. It is possible to prove (see [Mol90c]) that $\mathrm{mes}(W)^{1/2}(\hat{\lambda}_W - \lambda)$ converges in distribution to

$$\frac{d!}{b_d} \sum_{i=0}^{d} \binom{d}{i}(-1)^{d+i}\beta_i\,,$$

where $(\beta_0, \ldots, \beta_d)$ is a Gaussian centred random vector with the covariances

$$\mathbf{E}\beta_i\beta_j = \exp\{\lambda b_d \mathbf{E}(\min(i,j) + \xi)^d\} - 1, \quad i = 0, \ldots, d.$$

The estimator $\hat{\lambda}_W$ is biased because of the logarithm in (4.5), which is necessary to estimate Ψ_Ξ. Bounds for its bias are given in [Mol90c] (for a variant of this estimator obtained by independent realisations of Ξ).

Dilations by segments. Another choice of three compact sets in $\mathbf{R}^d$ (for general d) suitable for the intensity estimation has been suggested in [Sch92]. Let K_1 and

K_2 be two central symmetric segments with lengths l_1 and l_2 and orientations given by the unit vectors u_1 and u_2. Then

$$T_\Xi(K_i) = 1 - \exp\left\{ - (\lambda \bar{A} + 2\lambda \mathbf{E}V(\Xi_0, K_i)) \right\}, \quad i = 1, 2, \quad (5.17)$$

where $V(\cdot, \cdot)$ is the mixed volume defined by $\mathrm{mes}(K \oplus F) = \mathrm{mes}(K) + \mathrm{mes}(F) + 2V(K, F)$. For the parallelogram $K = K_1 \oplus K_2$ we get

$$T_\Xi(K) = 1 - \exp\left\{ - (\lambda \bar{A} + 2\lambda \mathbf{E}V(\Xi_0, K_1) + 2\lambda \mathbf{E}V(\Xi_0, K_2) + l_1 l_2 \lambda [u_1, u_2]) \right\}, \quad (5.18)$$

where $[u_1, u_2] = \sin \beta_{u_1 u_2}$ is the volume of the parallelogram constructed on u_1 and u_2, and $\beta_{u_1 u_2}$ is the angle between u_1 and u_2. Now (5.17) and (5.18) together with the basic equation $p = 1 - \exp\{-\lambda \bar{A}\}$ allow us to estimate λ, $\bar{A}$ and the expected mixed volumes $\mathbf{E}V(\Xi_0, K_i)$, $i = 1, 2$.

Note that the value $2l_i^{-1}V(\Xi_0, K_i)$ is the $(d - 1)$-dimensional measure of the projection of K_i onto the hyperplane orthogonal to u_i. Therefore, this method yields estimators of the expected $(d - 1)$-dimensional measures of the projections of the typical grain.

6

Estimation of Set-Valued Individual Parameters

Modern stochastic geometry and spatial statistics are interested not only in estimators of numerical parameters, but also in estimators of shape characteristics. For the Boolean model these shape characteristics can be represented through different set-valued parameters. Below *individual* set-valued parameters are considered, whereas Section 3.5 dealt with aggregate set-valued characteristics. Note that most of the introduced set-valued parameters are disks or balls if the Boolean model in question is isotropic. Thus, they are interesting mostly in the anisotropic case.

6.1   Mean Difference Body

The *difference body* of Ξ_0 is defined as

$$\check{\Xi}_0 = \Xi_0 \oplus \check{\Xi}_0 = \{x - y \colon x, y \in \Xi_0\} \ . \tag{6.1}$$

If the grain Ξ_0 is central symmetric, then $\check{\Xi}_0 = \Xi_0$, and, therefore, $\check{\Xi}_0 = 2\Xi_0$. In this case the difference body determines Ξ_0 uniquely, while, in general, many convex sets can share the same difference body $\check{\Xi}_0$. The set $\check{\Xi}_0/2$ is said to be the *central symmetrisation* of Ξ_0, see [Lei80].

The Aumann expectation, $\mathbf{E}\check{\Xi}_0$, (see Section 2.5) of the random set $\check{\Xi}_0$ is said to be the *mean difference body*. Clearly, $\mathbf{E}\check{\Xi}_0$ is a disk (or ball) if Ξ_0 is an isotropic random compact set. The following estimators of the mean difference body were suggested in [Mol96b].

The *linear contact distribution function* provides information sufficient to estimate the mean difference body. In the anisotropic case the linear contact distribution function $H_u(r)$ depends on the direction of u and is given by

$$H_u(r) = 1 - \exp\{-\lambda r \mathbf{E}h(\tilde{\Xi}_0, v)\}, \quad r \geq 0, \; v \perp u, \|v\| = 1,$$

for any unit vector u. This formula follows from (4.16), since the area of $\Xi_0 \oplus [o, ur]$ is equal to the sum of $\mathsf{A}(\Xi_0)$ and the width of Ξ_0 in the direction v orthogonal to u multiplied by r, see Figure 6.1. In turn, this width is the support function of $\tilde{\Xi}_0$ in the direction v, since $h(\tilde{\Xi}_0, v) = h(\Xi_0, v) + h(\Xi_0, -v)$. Definition (2.16) of the Aumann expectation yields

$$h(\mathbf{E}\tilde{\Xi}_0, v) = -\frac{1}{\lambda r} \log(1 - H_u(r)). \tag{6.2}$$

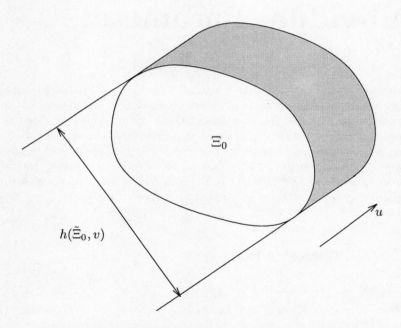

Figure 6.1 The width and the grain dilated by a segment.

Thus, (6.2) yields an estimator of the mean difference body:

$$\widehat{\mathbf{E}\tilde{\Xi}}_{0,W} \tag{6.3}$$
$$= \left\{ x: \langle x, v \rangle \leq -(r\hat{\lambda}_W)^{-1} \log(1 - \hat{H}_{u,W}(r)) \quad \text{for all } v, \|v\| = 1 \right\}.$$

This estimator depends on the choice of r. The value of r must not be very small, since it is impossible to choose freely the direction of a short vector on the digitised screen [MS94b, MSF93, Ser82]. On the other hand, larger values of r are not suitable because of considerable errors caused by the logarithmic transformation in (6.2), since $H_u(r)$ is close to 1 for large r. Estimator (6.3) is strong consistent with respect to the Hausdorff metric ρ_H, i.e.

$$\rho_H(\widehat{\mathbf{E}\tilde{\Xi}_{0,s}}, \mathbf{E}\tilde{\Xi}_0) \to 0 \quad \text{a.s. as} \quad s \to \infty.$$

Note that the Hausdorff distance between compact sets K and K_1 is given by

$$\rho_H(K, K_1) = \inf\{r \geq 0 : K \subseteq K_1^r, \ K_1 \subseteq K^r\},$$

see [Mat75, Mol95, Mol96b, SKM95].

Other estimators considered in [Mol96b] explore relationships between the mean difference body and set-valued aggregate parameters introduced in Section 3.5. For instance, the relationship between the mean difference body and the *mean star set* yields

$$h(\mathbf{E}\tilde{\Xi}_0, v) = (\lambda r(u))^{-1}, \quad \|u\| = \|v\| = 1, \ u \perp v, \tag{6.4}$$

where $r(u)$ is the radius-vector function of the mean star set S_Ξ, which is equal to the conditional expectation of the visibility extent given by (3.42). Note that (6.4) follows from the fact that the visibility extent, $\zeta_u(S_x)$, in any direction u has the distribution function $H_u(\cdot)$ under the condition $x \notin \Xi$. The function $r(u)$ can be estimated by

$$\hat{r}_W(u) = \frac{1}{\mathsf{N}(W \cap \mathbb{Z}^2)} \sum_{x \in (\mathbb{Z}^2 \cap W) \setminus \Xi} \zeta_u(S_x),$$

where $\mathbb{Z}^2$ is a grid of points in $\mathbf{R}^2$. It is possible to prove that

$$\sup_{\|u\|=1} |\hat{r}_W(u) - r(u)| \to 0 \quad \text{a.s. as} \quad W \uparrow \mathbf{R}^2,$$

if the mean difference body, $\mathbf{E}\tilde{\Xi}_0$, contains a neighbourhood of the origin, see [Mol96b]. Thus, the estimator of the mean star set yields an estimator of the support function of the mean difference body of the typical grain.

Also the *Steiner compact* of the fibre process $\partial\Xi$ (boundary of Ξ) can be related to the mean difference body through the formula

$$\mathbf{E}\tilde{\Xi}_0 = \frac{\omega_{\pi/2}\mathfrak{S}}{\lambda(1-p)}, \tag{6.5}$$

where $\omega_{\pi/2}\mathfrak{S}$ denotes the clockwise rotation of the Steiner compact $\mathfrak{S}$ to the angle $\pi/2$. Indeed, for each direction u, the rose of intersection $P_L(u)$ (see p. 52) is given by

$$P_L(u) = 2\lambda(u)(1-p),$$

where $\lambda(u)$ is the intensity of the point process obtained by the linear section $\ell_u \cap \Xi$, and ℓ_u is the line with direction u. On the one hand, the function $P_L(u)/2 = \lambda(u)(1 - p)$ is the support function of the Steiner compact $\mathfrak{S}$ of the fibre process $\partial\Xi$, see [SKM95, p. 283]. On the other hand, it follows from [Mat75, p. 144] that $\Xi \cap \ell_u$ has the intensity given by $\lambda(u) = \lambda h(\tilde{\Xi}_0, u)$, which yields (6.5).

For the estimation of the mean difference body using (6.3), (6.4) or (6.5) an estimator of the intensity is needed. For this, one of the estimators from Chapter 5 can be chosen. A comparative study of the estimators based on the formulae above can be found in [Mol96b]. Generally speaking, the estimator based on (6.5) has the best properties and also is less sensitive to edge effects.

Notes to Section 6.1

Higher dimensions. It should be noted that for higher dimensions the above described approaches no longer work. Fortunately, it is possible to estimate the mean difference body in $\mathbf{R}^d$ using $(d-1)$-dimensional sections of the Boolean model, see [Mol96b].

The theoretical background is provided by a result of Matheron [Mat75, p. 144], which states that, for any unit vector u, $\Xi \cap u^\perp$ is the Boolean model of intensity

$$\lambda(u) = \lambda \mathbf{E}\mu_1(\Pi_u \Xi_0),$$

where $u^\perp$ is the $(d-1)$-dimensional hyperplane orthogonal to u, $\mu_1(\cdot)$ is the 1-dimensional Lebesgue measure, and $\Pi_u \Xi_0$ is the projection of Ξ_0 in the direction of u. Then

$$\lambda(u) = \lambda \mathbf{E}h(\tilde{\Xi}_0, u),$$

i.e. the mean difference body can be found through intensities of the $(d-1)$-dimensional sections of the Boolean model Ξ.

Linear contact distribution functions in higher dimensions. Similarly to p. 91 we obtain

$$H_u(r) = 1 - \exp\{-\lambda r \mathbf{E}\mu_{d-1}(\Pi_v \Xi_0)\}, \quad r \geq 0, \, r \perp u, \, \|v\| = 1.$$

Thus, it is possible to estimate the mean value, $\mathbf{E}\mu_{d-1}(\Pi_v \Xi_0)$, of the $(d-1)$-dimensional Lebesgue measure of the projection of Ξ_0 onto the hyperplane orthogonal to u.

6.2   Convexification and the Mean Body

We have already seen that the mean difference body can be estimated by several elementary methods described in Section 6.1. However, the *mean body* $\bar{\Xi}_0 = \mathbf{E}\Xi_0$ of the *grain itself* is more interesting than the mean difference body. Below we will outline an approach based on relationships between the mean body and the specific convexification (one of the set-valued aggregate parameters). The main idea was suggested by Weil [Wei95].

The specific convexification (see Section 3.5) is related to the mean body of the grain. Consider the convexification $\mathsf{co}\,(\Xi \cap W)$ of the Boolean model Ξ inside the window W. Using the technique of [Wei95] it is possible to prove that the specific convexification converges almost surely in the Hausdorff metric to the set $\lambda \bar{\Xi}_0$, i.e.

$$\hat{Z}_W = \frac{\mathsf{co}\,(\Xi \cap W)}{A(W)} \to \lambda \bar{\Xi}_0 = \lambda \mathbf{E}\Xi_0 \quad \text{a.s. as} \quad W \uparrow \mathbf{R}^2. \tag{6.6}$$

Therefore, $\hat{Z}_W$ is an asymptotically unbiased estimator of the set $\lambda \bar{\Xi}_0$. A procedure to construct the convexification was described in Section 3.5. Other methods were suggested in [Wei95]. It is possible also to construct an unbiased estimator using independent realisations of the Boolean model and the boundary correction like (3.33), see [Sch92, Wei95]. It should be noted that (6.6) is valid also for any non-convex simply connected grain Ξ_0 (the set Ξ_0 is called *simply connected* if any closed path in Ξ_0 can be continuously deformed into a point remaining the whole time in Ξ_0).

It is easy to see that an estimator of λ together with (6.6) will give us an estimator of the mean body $\bar{\Xi}_0$ of the grain itself. Fortunately, this approach makes it possible to estimate also the intensity (even for non-convex grains!). For this, rewrite (5.9) to obtain

$$\chi_A = (1-p)(\lambda - A(\lambda\bar{\Xi}_0, -\lambda\bar{\Xi}_0)). \tag{6.7}$$

The mixed area $A(\lambda\bar{\Xi}_0, -\lambda\bar{\Xi}_0)$ can be estimated by $A(\hat{Z}_W, -\hat{Z}_W)$. The latter can be found through polygonal approximations of $\hat{Z}_W$, see [Wei95]: namely, if K is a polygon with the outer normals $u_1, \ldots, u_n$ and the corresponding lengths of the edges $l_1, \ldots, l_n$, then

$$A(K, -K) = \frac{1}{2} \sum_{i=1}^{n} l_i h(K, -u_i),$$

where $h(K, \cdot)$ is the support function of K. Hence, the intensity can be estimated by

$$\hat{\lambda}_W = \frac{\hat{\chi}_{A,W}}{1 - \hat{p}_W} + A(\hat{Z}_W, -\hat{Z}_W), \tag{6.8}$$

where $\hat{\chi}_{A,W}$ and $\hat{p}_W$ are estimators of the specific connectivity number and the area fraction.

Although the estimator (6.8) was constructed under very general assumptions (the grain Ξ_0 almost surely belongs to the convex ring and is simply connected), its evaluation may cause considerable difficulties because of complicated numerical evaluations of the convexification and mixed areas.

Properties of the points process of exposed tangent points make it possible to build another estimator for the mean body of the grain. The corresponding estimator will be mentioned in Section 7.4 in relation to a more general problem of the estimation of the grain's distribution.

Notes to Section 6.2

Additive extension of the support function. The proof of (6.6) is based on the extension theorem for the support functions [Wei90, Wei95, WW84].

The support function is additive, that is,

$$h(K \cup K', \cdot) + h(K \cap K', \cdot) = h(K, \cdot) + h(K', \cdot)$$

for convex K and K' such that $K \cup K'$ is also convex. By Groemer's theorem [Gro78], h has a unique additive extension onto the convex ring. The same is true for the centred support function

$$h^*(K, \cdot) = h(K, \cdot) - \langle s(K), \cdot \rangle ,$$

where

$$s(K) = \frac{1}{b_d} \int\limits_{S^{d-1}} u h(K, u) du$$

is the Steiner point of K. The additive extension of h^* onto the convex ring is also denoted by h^*, so that $h^*(K_1 \cup K_2) = h^*(K_1) + h^*(K_2) - h^*(K_1 \cap K_2)$. Then the spatial density of the extended centred support function satisfies

$$h_A^*(\cdot) = \lim_{W \uparrow \mathbf{R}^d} \frac{\mathbf{E} h^*(\Xi \cap W, \cdot)}{\mathsf{A}(W)} = \mathbf{E}\left[h^*(\Xi \cap W_0, \cdot) - h^*(\Xi \cap \partial^+ W_0, \cdot) \right], \qquad (6.9)$$

where W_0 is the unit square in $\mathbf{R}^2$ and $\partial^+ W_0$ is its upper-right boundary, see also p. 46.

The Boolean model assumption yields

$$\mathbf{E} h^*(\Xi \cap W, \cdot) = \mathsf{A}(W)(1-p)\lambda h^*(\mathbf{E}\Xi_0, \cdot) + h^*(W, \cdot)p$$

for all W from the convex ring. The evaluation of the density and integration over the unit circle yield

$$\mathsf{U}(\mathbf{E}\Xi_0) = \lambda^{-1} L_A ,$$

where L_A is the specific boundary length, cf. (3.45) and (6.5).

Note that (6.9) gives a way to estimate the mean body of the grain if independent identical realisations of Ξ within a unit cube are available.

General grains. If the typical grain is not simply connected but belongs to the convex ring, then (6.7) must be replaced by

$$\chi_A = (1-p)(\lambda \bar{\chi} - \lambda^2 \mathsf{A}(\lambda \mathbf{E}\Xi_0, -\lambda \mathbf{E}\Xi_0)) ,$$

where $\bar{\chi} = \mathbf{E}\chi(\Xi_0)$.

Higher dimensions and planar sections. The extension of the method based on extended support functions for higher dimensions is not straightforward. Moreover, the method as described in Section 6.2 does not work any more. A similar approach can be developed using surface area measures instead of support functions and the Blaschke expectation of Ξ_0 (see p. 54) instead of the Aumann expectation [Wei95].

7

Individual Parameters: Distributions

The numerical and set-valued parameters (and their estimators) are important to understand the nature of the Boolean model. However, these parameters do not completely characterise the *distribution* of the typical grain and, hence, the distribution of the Boolean model. The estimation of the typical grain's distribution is, clearly, the ultimate goal in statistical estimation for the Boolean model's parameters.

If the distribution of the grain belongs to a certain parametric family (see Section 8.1), then it is possible to find its parameters through some numerical mean values. For example, if the grain is a disk of random lognormally distributed radius ξ, then the distribution of ξ is retrievable by its first two moments, which are related to the mean perimeter and the mean area of the disk.

Note that all methods of Chapters 5 and 6 applied to the case of circular grains yield only estimates of the intensity, λ, and the two moments, $\mathbf{E}\xi$ and $\mathbf{E}\xi^2$, of the grain's radius ξ. This is, in general, not enough to retrieve the distribution of ξ. This chapter discusses several *non-parametric* estimators of the grain's distribution.

7.1 Deterministic Grain

If the grain Ξ_0 is *deterministic*, then the corresponding estimation problem can be considered to be parametric with a *set-valued* parameter. Since the mean body of a convex set coincides with the corresponding set, it is possible to use estimators of the mean body (or mean difference body) from Chapter 6 as estimators of the grain (or its difference body).

Another approach to estimating the difference body of the convex deterministic grain has been suggested in [Mol91a, Mol92, Mol94a]. It is based on the tail behaviour of the covariance function $C(\cdot)$ given by (2.17) or, equivalently, the function $q(\cdot)$ from (2.18). For this, it is necessary to find the shortest length of the vector v with varying direction such that $q(v)$ is exactly one. Unfortunately, the estimators of both $C(v)$ and $q(v)$ are unstable and fluctuate a lot if $\|v\|$ is large. To construct a stable statistical procedure let us put

$$\hat{\Xi}_{0,W}(\varepsilon) = \{x\colon \log \hat{q}_W(v) \geq \varepsilon\} \tag{7.1}$$

for some $\varepsilon > 0$, where $\hat{q}_W(v)$ is the empirical estimator of $q(v)$, see Section 4.2. It is possible to prove that the estimator $\hat{\Xi}_{0,W}(\varepsilon)$ converges almost surely in the Hausdorff metric to the set

$$\Xi_0(\varepsilon) = \{x\colon \lambda\mathsf{A}(\Xi_0 \cap (\Xi_0 - v)) \geq \varepsilon\}$$

as $W \uparrow \mathbf{R}^2$. In turn, the set $\Xi_0(\varepsilon)$ approximates the difference body $\tilde{\Xi}_0 = \Xi_0 \oplus \check{\Xi}_0$, that is to say

$$\left(1 - \left(\varepsilon/\lambda\mathsf{A}(\Xi_0)\right)^{1/d}\right)\tilde{\Xi}_0 \subset \Xi_0(\varepsilon) \subset \tilde{\Xi}_0. \tag{7.2}$$

Thus, (7.1) provides a kind of ε-biased estimator of the difference body $\tilde{\Xi}_0$. The bias can be removed by using of a sequence of ε's that converges to zero sufficiently slowly [Mol91a].

Estimator (7.1) depends on the covariance only, i.e. it uses the two-points sampling scheme. As a result, it is quite flexible and works in different observation schemes, where it is possible to estimate the covariance, see [Mol92, Mol94a] and Section 8.4. However, the bad tail behaviour of the covariance estimators makes it necessary to use larger values of ε, which increase the bias of the estimator (7.1).

Notes to Section 7.1

Properties of the estimator (7.1). The strong consistency of $\hat{\Xi}_{0,W}(\varepsilon)$ can be proved as follows, see [Mol94a]. First, note that $\Xi_0(\varepsilon + \delta) \subset \Xi_0(\varepsilon + \delta_1)$ if $\delta > \delta_1$. The next step is to prove that

$$\psi(\delta) = \rho_H(\Xi_0(\varepsilon), \Xi_0(\varepsilon + \delta)) \to 0 \quad \text{as} \quad \delta \to 0.$$

Then the monotonicity of $\Xi_0(\varepsilon)$ yields

$$\rho_H(\hat{\Xi}_{0,W}(\varepsilon) \cap K_0, \Xi_0(\varepsilon) \cap K_0) \leq \psi(3\nu_W(K_0 \cup \{o\})),$$

where

$$\nu_W(K_0 \cup \{o\}) = \sup_{x \in K_0 \cup \{o\}} \left|\hat{\Psi}_{\Xi,W}(\{o, x\}) - \Psi_{\Xi}(\{o, x\})\right| \to 0 \quad \text{a.s. as} \quad W \uparrow \mathbf{R}^d$$

by (4.6). Thus,

$$\rho_H(\hat{\Xi}_{0,W}(\varepsilon) \cap K_0, \Xi_0(\varepsilon) \cap K_0) \to 0 \quad \text{a.s. as} \quad W \uparrow \mathbf{R}^d$$

for each compact set K_0.

Furthermore, (7.2) follows from the fact that

$$\text{mes}(\Xi_0 \cap (\Xi_0 + x)) \geq (1 - g(x))^d \text{mes}(\Xi_0),$$

where $x \in \tilde{\Xi}_0$ and $g(x)$ is the *distance function* of $\tilde{\Xi}_0$, see [Lei80, Th. 20.3]. The distance function is determined by the properties

$$x = g(x)x_0, \quad x_0 \in \partial\tilde{\Xi}_0, \quad 0 < g(x) \leq 1, \quad x \in \tilde{\Xi}_0.$$

Then

$$\begin{aligned}
\Xi_0(\varepsilon) &\supset \{x \in \tilde{\Xi}_0 : \lambda(1 - g(x))^d \text{mes}(\Xi_0) \geq \varepsilon\} \\
&= \{x \in \tilde{\Xi}_0 : g(x) \leq 1 - (\varepsilon/\lambda\text{mes}(\Xi_0))^{1/d}\} \\
&= \left(1 - (\varepsilon/\lambda\text{mes}(\Xi_0))^{1/d}\right)\tilde{\Xi}_0.
\end{aligned}$$

Finite typical grains. An estimator of a non-convex deterministic grain $\Xi_0 = \{a_1, \ldots, a_m\}$ containing a finite number of points was constructed in [Mol91b]. Then Ξ is a particular case of the Neymann–Scott point process. Moreover, the random set Ξ is equal to a superposition of m Poisson point processes equal up to deterministic translations.

Isotropic rotations of a deterministic grain. Suppose that $\Xi_0 = \omega M$, i.e. the grain is obtained as a random isotropic rotation of a deterministic convex set M. Then the technique based on tails of the covariance allows us to construct a set-valued estimator of M, see [Mol93b]. For this, the tails of three-point coverings probabilities ought to be analysed.

This estimator can be constructed as follows. First, for a unit vector e and $r > 0$, we get

$$\begin{aligned}
-\log(1 - q(re)) &= \lambda\mathbf{E}\text{mes}(\omega M \cap (\omega M + re)) \\
&= \lambda\mathbf{E}\text{mes}(M \cap (M + re)) = \lambda\gamma_M(r),
\end{aligned}$$

where $\gamma_M(r)$ is the isotropised set-covariance function of M, see [SS94, p. 122]. Introduce the function ϕ_M as follows

$$\phi_M(o, x, y) = \mathbf{E}\text{mes}(\omega M \cap (\omega M + x) \cap (\omega M + y))$$
$$= \lambda^{-1}\Big[\Psi_\Xi(\{o, x, y\}) + 3\Psi_\Xi(\{o\}) - \Psi_\Xi(\{o, x\}) - \Psi_\Xi(\{o, y\}) - \Psi_\Xi(\{o, y - x\})\Big].$$

For $\varepsilon \in [0, 1]$ put

$$r(\varepsilon) = \sup\{r \geq 0 : \log(1 - q(re)) \leq \log(1 - p)\}.$$

Furthermore, define

$$M(\varepsilon, \delta) = \{x : \phi_M(o, r(\varepsilon)e, x) \geq \delta\varepsilon\}.$$

Then $r(\varepsilon) \uparrow d_M$ and $M(\varepsilon, \delta) \to M \oplus \Delta_0$ in the Hausdorff metric as $\varepsilon \downarrow 0$, where

$$d_M = \sup_{u \in \mathbf{S}^{d-1}} \{h(M, u) + h(M, -u)\} = h(M, e_0) + h(M, -e_0)$$

for some $e_0 \in \mathbf{S}^{d-1}$, and Δ_0 is the singleton $M \cap (M + d_M e_0)$. Thus, $M(\varepsilon, \delta)$ approximates M up to a shift for small ε. Then estimators of q and Ψ_Ξ can be plugged in to estimate $r(\varepsilon)$. Finally, its estimated value is used to find the empirical counterpart of the set $M(\varepsilon, \delta)$. The parameters ε and δ play the same role as the bias parameter ε in (7.1). The resulting estimator is strong consistent in the Hausdorff metric [Mol93b]. Unfortunately, this estimator depends on the tails of the three-point covariance, which makes it very unstable in applications.

7.2 Distribution of the Radius of the Spherical Grain

Suppose that the grain $\Xi_0 = B_\xi(0)$ is the disk of random radius ξ having the cumulative distribution function F_ξ. A non-parametric estimator for the function

$$G_\xi(r) = \int_0^r F_\xi(x)dx \tag{7.3}$$

has been suggested in [Mol90c]. For this, the values of the capacity functional T_Ξ or its logarithmic transformation Ψ_Ξ (see (4.4)) on the family of test sets that consists of disks $B_r(o)$, their boundaries (circles) $\partial B_r(o)$, and circles with centres $\{o\} \cup \partial B_r(o)$, for $r \geq 0$, must be analysed. Let us put

$$\hat{G}_{\xi,W}(r) = \frac{\hat{\Psi}_{\Xi,W}(B_{2r}(o)) - \hat{\Psi}_{\Xi,W}(\{o\} \cup \partial B_{2r}(o))}{4r\pi\hat{\lambda}_W}, \tag{7.4}$$

where $\hat{\Psi}_{\Xi,W}$ is defined in (4.5) and $\hat{\lambda}_W$ is a strong consistent estimator of λ (for example, any intensity estimator from Chapter 5). Then $\hat{G}_{\xi,W}(r)$ is a uniformly strong consistent estimator of $G_\xi(r)$ for $r \in [a, b]$, i.e.

$$\sup_{a \leq r \leq b} \left| G_\xi(r) - \hat{G}_{\xi,W}(r) \right| \to 0 \quad \text{a.s. as} \quad W \uparrow \mathbf{R}^2$$

for all positive a and b. A central limit theorem for $\hat{G}_{\xi,W}$ is given in [Mol90c].

The same approach is applicable if $\Xi_0 = \xi M$, i.e. if the grain is a random scale transform of a deterministic central symmetric set M, see [Mol94b]. An analogue of (7.4) can be constructed for the sets rM and $\{o\} \cup r\partial M$ instead of $B_r(o)$ and $\{o\} \cup \partial B_r(o)$ respectively.

Properties of estimator (7.4) depend on the occurrences of separate grains or clumps with diameters less than the given value of r. Thus, for higher area fractions, estimator (7.4) is bad. In general, higher area fractions require larger windows of observations.

Another approach to estimating the distribution of the radius of the spherical grain was suggested by Hall [Hal88, pp. 321–323]. It is based on curvature measurements at boundary points of Ξ.

Let f be a certain numerical function on $(0, \infty)$. Hall considered statistics of the type

$$\kappa_W(f) = \mathsf{A}(W)^{-1} \sum_{i=1}^{\infty} \phi_i f(R_i),\tag{7.5}$$

where the sum stretches over all protruding pieces of the grain boundaries within the window W, R_i is the radius and ϕ_i is the angular content of the ith protruding piece. Note that several protruding pieces may belong to the same grain. The measurement technique was discussed in [Hal88, pp. 322, 323].

If $\mathbf{E}|f(\xi)| < \infty$, then $\mathbf{E}f(\xi)$ can be estimated, since

$$\mathbf{E}\kappa_W(f) = 2\pi\lambda \exp\{-\lambda\pi\mathbf{E}\xi^2\}\mathbf{E}f(\xi) = 2\pi\lambda(1-p)\mathbf{E}f(\xi).\tag{7.6}$$

Thus, estimators of λ and p can be plugged in to get an estimator of $\mathbf{E}f(\xi)$. In particular, the moments of ξ can be estimated. For instance,

$$\mathbf{E}\kappa_W(1) = 2\pi\lambda(1-p),$$

whence

$$\hat{m}_{f(\xi), W} = \frac{\kappa_W(f)}{\kappa_W(1)}$$

is a ratio-unbiased and strong consistent estimator of $\mathbf{E}f(\xi)$.

The limit theory does not differ crucially from the case of independent observations, since curvatures for different grains are independent (although some care must be taken with respect to protruding pieces belonging to the same grain).

A method of estimating the density of ξ from the angular intensity distribution (small-angle scattering experiment) has been suggested in [Gil95]. It is based on the function

$$\gamma(r) = \frac{C(r) - p^2}{p(1-p)}$$

which can be computed from the scattering intensity (because of isotropy, $C(r)$ is the value of the covariance function at the vector with norm r). In $\mathbf{R}^3$, the second derivative of the function

$$t(r) = \frac{1}{\lambda\mathbf{E}\mu_3(B_\xi)} \log\left[\frac{\gamma(r)p}{1-p} + 1\right] = \frac{\log q(r)}{\lambda\mathbf{E}\mu_3(B_\xi)},$$

can be used to find the density of ξ, see (2.18).

Furthermore, it is possible to estimate the density of ξ by means of the empirical covariance using the so-called *regularisation technique*, see Section 8.3.

Notes to Section 7.2

Function G_ξ and the capacity functional. By (4.4), we get in the d-dimensional case

$$\Psi_\Xi(B_r) = \lambda b_d \mathbf{E}(\xi + r)^d = \lambda b_d \int_0^\infty (x + r)^d dF_\xi(x),$$

$$\Psi_\Xi(\partial B_r) = \Psi_\Xi(B_r) - \lambda b_d \int_0^r (r - x)^d dF_\xi(x),$$

$$\Psi_\Xi(\partial B_{2r} \cup \{o\}) = \Psi_\Xi(B_{2r}) - \lambda b_d \int_0^r \left((2r - x)^d - x^d\right) dF_\xi(x).$$

Using these formulae one can relate the function $G_\xi(\cdot)$ from (7.3) to the functional $\Psi_\Xi(\cdot)$. For $d \leq 3$ we get

$$G_\xi(r) = \begin{cases} (\Psi_\Xi(B_{2r}) - \Psi_\Xi(\partial B_{2r} \cup \{o\}))/(4\lambda) & , d = 1, \\ (\Psi_\Xi(B_{2r}) - \Psi_\Xi(\partial B_{2r} \cup \{o\}))/(4r\lambda\pi) & , d = 2, \\ (\Psi_\Xi(B_{2r}) - \Psi_\Xi(\partial B_{2r} \cup \{o\}) - 2\Psi_\Xi(B_r) \\ \qquad + 2\Psi_\Xi(\partial B_r))/(8r^2\lambda\pi) & , d = 3. \end{cases}$$

Similarly to (7.4), one can use these formulae to build plug-in estimators of $G_\xi(r)$.

Limit properties of statistics (7.5). Formula (7.6) and a central limit theorem for statistics $\kappa_W(f)$ can be obtained from general results of [HM95]. They are valid in any dimension d and even for more general sums of weighted curvature measures and general smooth convex grains. In particular, if $\mathbf{E}\xi^4$ and $\mathbf{E}f(\xi)^2$ are finite, then $\mathrm{mes}(W)^{1/2}(\kappa_W(f) - \mathbf{E}\kappa_W(f))$ converges weakly to a Gaussian centred random variable with the variance

$$\sigma^2_{\kappa(f)} = \lambda(1-p)^2 \mathbf{E}\left[\left(\frac{f(\xi)}{\xi}\right)^2 \int_{\partial\Xi_0} q(y_1 - y_2) dy_1 dy_2\right]$$

$$+ \lambda^2(1-p)^2 \int_{\mathbf{R}^d} \left[q(x)b(x)b(-x) - 4\pi^2(\mathbf{E}f(\xi))^2\right] dx,$$

where $q(\cdot)$ is defined in (2.18), the integration over $\partial\Xi_0$ is understood with respect to the $(d-1)$-dimensional Hausdorff measure, and

$$b(x) = \mathbf{E}\left[\frac{f(\xi)}{\xi} \int\limits_{\partial\Xi_0 \cap (\Xi_0^c + x)} dy\right]$$

is the expected angular content of $\partial\Xi_0$ within $\Xi_0^c + x$ multiplied by $f(\xi)$.

7.3   Covering Probabilities

The point process $N^+(u)$ of (positive) tangent points in the direction u was introduced in Section 3.2. Its intensity, N_A^+, was used in Section 5.3 to estimate the intensity λ of the Boolean model. Below we will exploit further information contained in this process to estimate the grain's distribution.

The second-order properties of the point process $N^+(u)$ are summarised in its pair-correlation function $g_u(v)$, see Section 4.4 for the definition and the estimation technique. Note that in Section 5.3 this function has been used to find the asymptotic variance of the intensity estimator. Below we will see that this pair-correlation function $g_u(\cdot)$ is itself useful when we are estimating the distribution of the grain.

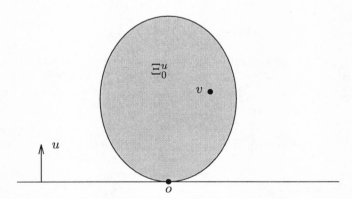

Figure 7.1   Shifted grain $\Xi_0^u = \Xi_0 - n_u(\Xi_0)$.

It was proved in [MS94a] (see also p. 38) that

$$g_u(v) = q(v)\psi_u(v)\psi_u(-v)\,, \tag{7.7}$$

where q is defined by (2.18) and $\psi_u(v) = \mathbf{P}\{v \notin \Xi_0^u\}$ is related to the covering probabilities of the shifted grain, see Figure 7.1. For the shifted grain Ξ_0^u

the origin serves as the corresponding tangent point in the direction u (u-tangent point). Thus, the grain is a subset of the half-plane with the boundary orthogonal to u. If v and u are not orthogonal, then either v or $(-v)$ lies in the other ('lower') half-plane, whence the corresponding value of ψ_u is exactly 1. For example, if u is directed upwards, then $\psi_u(-v) = 1$ and

$$g_u(v) = q(v)\psi_u(v)$$

for all v with positive ordinates. In general,

$$\psi_u(v) = \mathbf{P}\{v \notin \Xi_0^u\} = \begin{cases} g_u(v)/q(v) & , \quad \langle u, v \rangle \geq 0, \\ 1 & , \quad \text{otherwise}, \end{cases} \tag{7.8}$$

where $\langle u, v \rangle$ is the scalar product of u and v. Thus, the *covering probabilities*

$$t_u(v) = 1 - \psi_u(v) = \mathbf{P}\{v \in \Xi_0^u\}$$

can be estimated by

$$\hat{t}_{u,W}(v) = 1 - \frac{\hat{g}_{u,W}(v)}{\hat{q}_W(v)}, \quad \langle u, v \rangle \geq 0, \tag{7.9}$$

using the estimators $\hat{q}$ and $\hat{g}_{u,W}$ of the functions q and g_u (see Sections 4.2 and 4.4). Clearly, $\hat{t}_{u,W}(v) = 0$ for all v such that $\langle u, v \rangle < 0$. From these covering probabilities it is possible to estimate the median

$$\text{Med}\,\Xi_0^u = \{v\colon t_u(v) \geq 1/2\}$$

(see [SS94, p. 115]) or quantiles [Mol90b] of the random set Ξ_0^u.

Furthermore, (7.8) allows us to estimate the distribution of the special typical grain $\Xi_0 = B_\xi(o)$, which is the random disk of radius ξ. Indeed, then

$$t_u(v) = \begin{cases} \mathbf{P}\left\{\xi \geq \frac{\|v\|^2}{2(u \cdot v)}\right\} & , \quad \langle u, v \rangle > 0, \\ 1 & , \quad v = o, \\ 0 & , \quad \text{otherwise}, \end{cases}$$

see [Mol95]. In particular,

$$t_u(v) = \mathbf{P}\{\xi \geq r/2\}, \quad r \geq 0, \, v = ru. \tag{7.10}$$

If u is directed upwards, then the corresponding point $v = ru = (0, r)$ lies on the ordinate axis. Therefore, the covering probabilities of the shifted grain yield the distribution function of the radius of the spherical grain. In contrast to Section 7.2, here the distribution function itself rather than its integral is estimated.

Further finite-points *hitting probabilities*

$$t_u(v_1, \ldots, v_n) = \mathbf{P}\{\{v_1, \ldots, v_n\} \cap \Xi_0^u \neq \emptyset\} \qquad (7.11)$$

can be estimated from higher-order product densities of the point process $N^+(u)$ [Mol95, MS94a]. These hitting probabilities determine uniquely the distribution of the typical grain Ξ_0. For instance, the capacity functional $T_{\Xi_0}(K)$ of Ξ_0 can be retrieved by using finite-point approximations of K, see [Mat75, Section 2.4]. Thus, the information contained in the point process $N^+(u)$ for one u together with the covariance function (or the function q) is sufficient to estimate the intensity and the grain's distribution, i.e. *all* parameters of the Boolean model. Unfortunately, the practical implementation of this approach is difficult because of bad quality of estimators of higher-order product densities.

By integration of covering probabilities it is possible to evaluate moments of the grain's area. In the simplest case,

$$\int_{\mathbf{R}^2} t_u(v)dv = \int_{\mathbf{R}^2} \left(1 - \frac{g_u(v)}{q(v)}\right) dv = \mathbf{EA}(\Xi_0).$$

Integration over compact K yields the expectation of the area of the intersection of Ξ_0^u and K (see p. 13),

$$\int_K t_u(v)dv = \mathbf{EA}(\Xi_0^u \cap K).$$

Note that it is impossible to replace Ξ_0^u by Ξ_0 in the last equation.

Furthermore, the two-points covering probabilities can be integrated to get

$$\int_{\mathbf{R}^2}\int_{\mathbf{R}^2} \mathbf{P}\{\{v_1, v_2\} \subset \Xi_0^u\}\, dv_1 dv_2 = \mathbf{EA}(\Xi_0)^2.$$

Since

$$\mathbf{P}\{\{v_1, v_2\} \subset \Xi_0^u\} = t_u(v_1) + t_u(v_2) - t_u(v_1, v_2),$$

we obtain

$$\int_{\mathbf{R}^2}\int_{\mathbf{R}^2} t_u(v_1, v_2)dv_1 dv_2 = 2\mathbf{EA}(\Xi_0) - \mathbf{EA}(\Xi_0)^2.$$

Notes to Section 7.3

Quantiles of random sets. A general approach to defining quantiles of random sets was elaborated in [Mol90b]. Let X be a random closed set in $\mathbf{R}^d$, and let $\mathfrak{M}$ be a subfamily of $\mathcal{K}$. Then the pth quantile of X is defined by

$$M_p = \bigcup\{K \in \mathfrak{M} : T_X(K) < p\}.$$

This set can be estimated if an estimator of $T_X(K)$, $K \in \mathfrak{M}$, (empirical capacity functional) is available. The strong consistency and a central limit theorem for the corresponding estimator of M_p can be found in [Mol90b].

For simplicity, let $\mathfrak{M}$ be the family of all singletons belonging to a fixed compact set K_0. Put $t(x) = \mathbf{P}\{x \in X\}$, $x \in K_0$. Then

$$M_p = \{x \in \mathbf{R}^d : t(x) < p\}.$$

In particular, $\mathbf{R}^d \setminus M_{1/2}$ is the *median* of X, see [SS94, p. 115]. Assume that $\hat{t}_n(x)$ is a uniformly strong consistent estimator of $t(x)$, i.e.

$$\sup_{x \in K_0} |\hat{t}_n(x) - t(x)| \to 0 \quad \text{a.s. as} \quad n \to \infty.$$

For instance, $\hat{t}_n$ can be taken from (7.9) for a sequence of windows $W = W_n \uparrow \mathbf{R}^d$. Put $\hat{M}_{p,n} = \{x \in \mathbf{R}^d : \hat{t}_n < p\}$.

If $M_{p+} = \{x : t(x) \leq p\}$ is contained in the topological closure of M_p, then

$$\rho_H(\hat{M}_{p,n}, M_p) \to 0 \quad \text{a.s. as} \quad n \to \infty,$$

i.e. $\hat{M}_{p,n}$ is a strong consistent estimator of the pth quantile M_p.

Furthermore, assume that $a_n(\hat{t}_n(x) - t(x))$ satisfies the functional limit theorem in the space $C(K_0)$ of continuous functions on K_0. Its weak limit as $a_n \to \infty$ is denoted by $\zeta(x)$. Define

$$\omega(x; \delta) = \inf\{t(y) - t(x) : y \in K_0, \ \rho(x, y) \leq \delta\}, \quad x \in K_0,$$

and suppose that $\mathbf{P}\{x \in \partial X\} = 0$ for all x and the function $\omega(x; \delta)$ is differentiable at $\delta = 0$ uniformly for $x \in \{x \in K_0 : |t(x) - p| \leq \varepsilon\}$ with upper semi-continuous non-vanishing derivative $L(x) = \omega'_\delta(x; 0)$. Then $a_n \rho_H(\hat{M}_{p,n}, M_p)$ converges in distribution to the random variable

$$\sup_{x \in K_0, \ t(x) = p} \left| \frac{\zeta(x)}{L(x)} \right|.$$

This result can be applied to the empirical covering probabilities of the grain in order to obtain asymptotic properties of the corresponding quantile estimators.

Higher-order product densities and covering probabilities. Let us describe an estimation procedure for the hitting probabilities defined in (7.11). Without loss of generality assume that points $v_1, \ldots, v_n$ belong to the half-space

$$H_u^+ = \{v \in \mathbf{R}^d : \langle v, u \rangle \geq 0\}$$

and are ordered according to the growing sequence $\langle v_i, u \rangle$, $1 \leq i \leq n$. Note that the product term in formula (3.18), which gives the higher-order product density $\rho_u^{(n)}(x_1, \ldots, x_n)$ of the point process $N^+(u)$, can be written as

$$\prod_{j=1}^n \psi_u(v_1 - v_j, \ldots, v_{j-1} - v_j, v_{j+1} - v_j, \ldots, v_n - v_j)$$

$$= \psi_u(v_n - v_{n-1})\psi_u(v_n - v_{n-2}, v_{n-1} - v_{n-2}) \cdots \psi_u(v_n - v_1, \ldots, v_2 - v_1)$$

with

$$\psi_u(v_1, \ldots, v_k) = 1 - t_u(v_1, \ldots, v_k).$$

The hitting probabilities from (7.11) can be estimated recurrently, since, by (3.18),

$$\frac{\rho_u^{(n+1)}(0, v_1, \ldots, v_n)}{\lambda^{n+1} Q(\{0, v_1, \ldots, v_n\})} = \psi_u(v_n - v_{n-1})$$

$$\times \psi_u(v_n - v_{n-2}, v_{n-1} - v_{n-2}) \cdots \psi_u(v_n - v_1, \ldots, v_2 - v_1)\psi_u(v_1, \ldots, v_n).$$

In particular, for $n = 2$,

$$t_u(v_1, v_2) = 1 - \psi_u(v_1, v_2) = 1 - \frac{1}{\lambda} \frac{\rho_u^{(3)}(0, v_1, v_2)}{\rho_u^{(2)}(0, v_2 - v_1)} \frac{Q(\{0, v_2 - v_1\})}{Q(\{0, v_1, v_2\})}.$$

Capacity functional of Ξ_0. Hitting probabilities (7.11) determine uniquely the capacity functional and, thereupon, the distribution of the shifted grain Ξ_0^u. Indeed, since Ξ_0 is regular closed (a.s. coincides with the closure of its interior), the covering probability for compact set K can be found by finite-points approximations:

$$\mathbf{P}\{K \subset \Xi_0^u\} = \sup\{\mathbf{P}\{\{v_1, \ldots, v_n\} \subset \Xi_0^u\} : v_1, \ldots, v_n \in K, n \geq 1\}.$$

In turn, $\mathbf{P}\{\{v_1, \ldots, v_n\} \subset \Xi_0^u\}$ can be found via hitting probabilities $t_u(v_1, \ldots, v_n)$ by the inclusion–exclusion formula.

Integrals of $\mathbf{P}\{\{v_1, \ldots, v_n\} \subset \Xi_0^u\}$ determine moments of the volume of the typical grain, i.e.

$$\int_{H_u^+} \cdots \int_{H_u^+} \mathbf{P}\{\{v_1, \ldots, v_n\} \subset \Xi_0^u\} \, dv_1 \cdots dv_n = \mathbf{E}[\mathrm{mes}(\Xi_0)^n], \quad n \geq 1.$$

7.4   Finite-Dimensional Distributions of the Support Function

In this section we will deal with joint distributions of point processes of tangent points defined for different directions. The approach is based on the evaluation of the pair-correlation function $g_{u,w}(v)$ of the point processes $N^+(u)$ and $N^+(w)$ of tangent points in the directions u and w, see Section 4.4. It was shown in [Mol95] (see also p. 39) that

$$g_{u,w}(v) = q(v)\psi_u(v)\psi_w(-v) + \begin{cases} \lambda^{-1} p_{u,w}(v)q(v) & , \quad u \neq w, \\ 0 & , \quad u = w. \end{cases} \quad (7.12)$$

Note that $(N_A^+)^2 g_{u,w}(v)$ is the infinitesimal probability of having a u-tangent point in the neighbourhood of the origin and a w-tangent point in the neighbourhood of v. The functions $q(\cdot)$ and $\psi_u(\cdot)$ have the same meanings as above, and $p_{u,w}(\cdot)$ is the probability density function of the random vector $\xi_{u,w}$ that joins the tangent points of Ξ_0 in the directions u and w, see Figure 7.2. As in Section 7.3, we abbreviate $g_{u,u}(v)$ as $g_u(v)$.

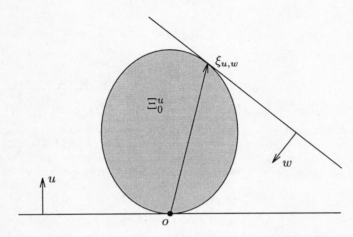

Figure 7.2 Two tangent points and the random chord $\xi_{u,w}$.

The probability density function $p_{u,w}(\cdot)$ can be found from (7.7) and (7.12) as

$$p_{u,w}(v) = \frac{\lambda}{q(v)^2}\Big(g_{u,w}(v)q(v) - g_u(v)g_w(v)\Big).\qquad(7.13)$$

Therefore, $p_{u,w}(v)$ can be estimated using already known estimators of functions and parameters in the right-hand side. Properties of the corresponding estimators depend very much on the properties of estimators of the pair-correlation functions $g_{u,w}(\cdot)$ and $g_u(\cdot)$.

In some special cases the function $p_{u,w}(\cdot)$ yields the distribution of the grain. For example, if $\Xi_0 = B_\xi(o)$, then $p_{u,-u}(v)$ is the *density* of the random variable 2ξ.

In many other cases it is necessary to work with three point processes of tangent points, $N^+(u_i)$, $i = 1, 2, 3$, and the corresponding joint product density $p_{u_1,u_2,u_3}(v_1, v_2)$. This allows us to estimate the distribution of the axes of an isotropic elliptical grain, the joint distributions of sides of a rectangular grain or the length of a 'long thin' grain, see [Mol95] and p. 113. The typical choice of the directions is $u_1 = u$, $u_2 = -u$ and $u_3 = u^\perp$ for some u.

The probability density function $p_{u,w}(v)$ yields the *distribution of the support function* of Ξ_0^u. Note that

$$h(\Xi_0^u, w) = \langle \xi_{u,w}, w \rangle,\qquad(7.14)$$

whence the cumulative distribution function of $h(\Xi_0^u, w)$ can be found as

$$\mathbf{P}\left\{h(\Xi_0^u, w) \leq r\right\} = \int\limits_{\{v:\, 0 \leq \langle u,v \rangle \leq r\}} p_{u,w}(v)dv\,.$$

The expectation of (7.14) yields the support function of the *mean body* of the shifted grain, i.e.

$$h(\mathbf{E}\Xi_0^u, w) = \mathbf{E}h(\Xi_0^u, w) = \int\limits_{\{v:\, \langle u,v \rangle \geq 0\}} \langle u,v \rangle p_{u,w}(v)dv\,.$$

Clearly, $\mathbf{E}\Xi_0^u$ is equal to $\mathbf{E}\Xi_0$ up to a translation.

Furthermore, the scalar product $\langle \xi_{u,-u}, u \rangle$ is equal to the width

$$b(\Xi_0, u) = h(\Xi_0, u) + h(\Xi_0, -u)$$

of the grain in the direction u. The expected width gives the support function of the mean difference body, see p. 93. Thus,

$$\mathbf{E}h(\tilde{\Xi}_0, u) = \mathbf{E}b(\Xi_0, u) = \int\limits_{\langle u,v \rangle \geq 0} \langle u,v \rangle p_{u,-u}(v)dv\,.$$

In general, the point processes $N^+(u_i)$, $1 \leq i \leq m$, of tangent points in different directions yield estimators of *finite-dimensional distributions* of the support function $h(\Xi_0^u, v)$, $\|v\| = 1$, as a random function on the unit circle [Mol95].

Notes to Section 7.4

Tangent points in three directions. Higher-order characteristics of the marked point process $N^+(u_1, \ldots, u_m)$ (see p. 38) of tangent points in different directions can be expressed by means of the functions $\psi_{u_i}(\cdot)$ from (3.19) and the joint distribution $P_{u_1,\ldots,u_m}(\cdot)$ of the random vector $(\xi_{u_1,u_2}, \ldots, \xi_{u_1,u_m})$. We consider below only the particular case of three directions ($m = 3$).

Theorem 7.1 (see [Mol95]) *Assume that the distribution P_{u_1,u_2,u_3} of $(\xi_{u_1,u_2}, \xi_{u_1,u_3})$ has the density $p_{u_1,u_2,u_3}(\cdot)$ with respect to Lebesgue measure, and the distributions P_{u_i,u_j} of ξ_{u_i,u_j} have densitites $p_{u_i,u_j}(\cdot)$, $i,j = 1,2,3$, $i \neq j$. Then the moment measure*

$$\mu_{u_1,u_2,u_3}^{(3)}(W_1 \times W_2 \times W_3) = \mathbf{E}\left[N^+(u_1, W_1)N^+(u_2, W_2)N^+(u_3, W_3)\right]$$

has the density with respect to Lebesgue measure given by

$$\rho_{u_1,u_2,u_3}^{(3)}(x, y, z)$$

$$= Q(\{x, y, z\}) \Big[\lambda^3 \psi_{u_1}(y - x, z - x) \psi_{u_2}(x - y, z - y) \psi_{u_3}(x - z, y - z)$$

$$+ \lambda^2 \Big(p_{u_1, u_2}(y - x) \psi_{u_1}(z - x) \psi_{u_3}(x - z, y - z)$$

$$+ p_{u_1, u_3}(z - x) \psi_{u_1}(y - x) \psi_{u_2}(x - y, z - y)$$

$$+ p_{u_2, u_3}(z - y) \psi_{u_3}(y - z) \psi_{u_1}(y - x, z - x) \Big) + \lambda p_{u_1, u_2, u_3}(y - x, z - x) \Big] .$$

This result allows us to estimate the density $p_{u_1, u_2, u_3}(v_1, v_2)$, since estimators for all other functions in the right-hand side are already known. The left-hand side can be estimated by means of a kernel estimator, see [SS94, p. 294]. Below we will consider particular cases of this approach for special Boolean models.

Elliptical grains. The case of spherical grain was discussed in Section 7.3. Remember that the distribution of the radius is related to the covering probabilities by (7.10) and can be estimated using the point process of tangent points in one direction.

Let us consider the isotropic planar Boolean model with the grain Ξ_0 being equal to the ellipse with random half-axes ζ_1 and ζ_2. Furthermore, let $\eta = (\eta_1, \eta_2)$ be an isotropic unit random vector (the orientation of the longer axis of the ellipse). Then, for any $u \in \mathbf{S}^{d-1}$,

$$\left(\sqrt{\zeta_1^2 \eta_1^2 + \zeta_2^2 \eta_2^2}, \sqrt{\zeta_1^2 \eta_2^2 + \zeta_2^2 \eta_1^2} \right) \overset{d}{\sim} \frac{1}{2} \left(b(\Xi_0, u), b(\Xi_0, u^\perp) \right) ,$$

whence

$$\zeta_1^2 + \zeta_2^2 \overset{d}{\sim} \frac{1}{4} \left(b(\Xi_0, u)^2 + b(\Xi_0, u^\perp)^2 \right)$$

($\overset{d}{\sim}$ means equivalence of distributions). Thus, the distribution of $\zeta_1^2 + \zeta_2^2$ can be evaluated via the distribution of the random vector $(b(\Xi_0, u), b(\Xi_0, u^\perp))$.

Furthermore, elementary (but lengthy) calculations yield

$$\left(\frac{\zeta_1^2 \eta_1}{\sqrt{\zeta_1^2 \eta_1^2 + \zeta_2^2 \eta_2^2}}, \frac{\zeta_2^2 \eta_2}{\sqrt{\zeta_1^2 \eta_1^2 + \zeta_2^2 \eta_2^2}} \right) \overset{d}{\sim} -\frac{1}{2} \xi_{u, -u}$$

and

$$\left(\frac{-\zeta_1^2 \eta_2}{\sqrt{\zeta_1^2 \eta_2^2 + \zeta_2^2 \eta_1^2}}, \frac{\zeta_2^2 \eta_1}{\sqrt{\zeta_1^2 \eta_2^2 + \zeta_2^2 \eta_1^2}} \right) \overset{d}{\sim} \xi_{u, u^\perp} - \frac{1}{2} \xi_{u, -u} .$$

Hence,

$$\frac{\zeta_1^4}{\zeta_2^4} \overset{d}{\sim} -\frac{\xi_{u, -u}^{(1)}}{\xi_{u, -u}^{(2)}} \frac{\xi_{u, u^\perp}^{(1)} - \frac{1}{2} \xi_{u, -u}^{(1)}}{\xi_{u, u^\perp}^{(2)} - \frac{1}{2} \xi_{u, -u}^{(2)}} ,$$

where $(\xi_{u, -u}^{(1)}, \xi_{u, -u}^{(2)})$ are components of the vector $\xi_{u, -u}$ etc. Note that the right-hand side is always positive. Therefore, the distribution of ζ_1 / ζ_2 can be obtained from the density $p_{u, u^\perp, -u}$.

Rectangular grains. Let the planar Boolean model Ξ have a rectangular typical grain with the sides of lengths ζ_1 and ζ_2 parallel to the coordinate axes. Then, for the vector $u = (0, 1)$ directed upwards,

$$t_u(v) = \mathbf{P}\{v_1 \leq \zeta_1, v_2 \leq \zeta_2\}, \quad v = (v_1, v_2).$$

Thus, the common distribution of the random variables ζ_1 and ζ_2 is simply related to the covering probabilities.

If the rectangle is isotropic, then $(\|\xi_{u,u^\perp}\|, \|\xi_{u,-u^\perp}\|)$ coincides in distribution with (ζ_1, ζ_2) (here $\|\xi\|$'s and ζ's have been arranged in descending order). Thus, the mutual distribution of (ζ_1, ζ_2) can be expressed through the density $p_{u,u^\perp,-u^\perp}(\cdot)$. For this, the moment measures of the marked point process $N(u, u^\perp, -u^\perp)$ up to the third order are necessary.

Long thin grains. Suppose that the grain Ξ_0 is a very thin rectangle (in the planar case) or prism (for $d \geq 3$). Such a case appears in the studies of the microstructure of paper, see [MSF93] and references therein. For this model the *distribution* of the length ζ of the grain is of interest. Fortunately, for almost all directions u the norm $\|\xi_{u,-u}\|$ is equal to the length of the diagonal of this rectangular grain. The latter is, in turn, approximately equal to ζ. Therefore, the distribution function of ζ can be found as follows:

$$\mathbf{P}\{\zeta \leq x\} \approx \mathbf{P}\{\|\xi_{u,-u}\| \leq x\} = \int_{B_x(0) \cap H_u^+} p_{u,-u}(v) dv.$$

The direction u must be a.s. non-orthogonal to the sides of the grain. In the isotropic case all directions will do.

Finite-dimensional distributions of the support function. Since $h(\Xi_0^u, v) = \langle \xi_{u,v}, v \rangle$, we get

$$\mathbf{P}\{h(\Xi_0^u, v_1) < c_1, h(\Xi_0^u, v_2) < c_2\} = \int_{\substack{0 \leq \langle x_1, u \rangle \leq c_1 \\ 0 \leq \langle x_2, u \rangle \leq c_2}} p_{u,v_1,v_2}(x_1, x_2) dx_1 dx_2.$$

Thus, an estimator of $p_{u,v_1,v_2}(x_1, x_2)$ can be used to estimate the joint cumulative distribution function of $h(\Xi_0^u, v_1)$ and $h(\Xi_0^u, v_2)$. Further finite-dimensional distributions of $h(\Xi_0, \cdot)$ can be expressed via densities $p_{u,v_1,\ldots,v_n}(x_1, \ldots, x_n)$ of the random vector $(\xi_{u,v_1}, \ldots, \xi_{u,v_n})$.

8

Other Sampling Schemes

8.1 Parametric Models and the Parametric Approach

Sometimes it is reasonable to assume that the distribution of the grain belongs to a parametric family. Then it is possible to estimate the distribution through estimators of numerical parameters without addressing the non-parametric technique described in Chapter 7.

Clearly, when working with parametric families it suffices to exploit the number of functionally independent aggregate parameters, which is equal to the number of parameters of the parametric family plus 1 (for the intensity λ).

The major part of the relevant studies assumes that the grain is a *disk* with random radius ξ having distribution belonging to a parametric family. The typical choices are listed below together with references to papers where the corresponding parametric families were applied in the Boolean model statistics. The density of the grain's radius ξ is denoted by $f(\cdot)$.

- The Gaussian distribution for the radius of the grain was considered in [Dup80] and [Hal88, pp. 318–321]. To exclude negative values it is necessary to assume that the expectation, μ, is greater than 3σ for the standard deviation, σ.
- The shifted Weibull distribution with the density

$$f(r) = \begin{cases} 0 & , \ r < \delta\,, \\ k\rho(r-\delta)^{k-1}\exp\{-\rho(r-\delta)^k\} & , \ r \geq \delta\,, \end{cases} \tag{8.1}$$

which depends on three parameters k, ρ and δ, see [Dig81].
- The lognormal distribution with the density

$$f(r) = \frac{1}{\sqrt{2\pi}\sigma r}\exp\left\{-\frac{(\log r - \mu)^2}{2\sigma^2}\right\}\,, \tag{8.2}$$

which depends on two parameters μ and σ, see [BS91].

Below we will consider the estimation technique suggested by Dupač [Dup80] for the case of normally distributed radii, see also [Hal88, Section 5.5]. This method is based on properties of circular clumps.

A circular clump is a connected part of Ξ that is a perfect circle. Such a clump can appear only if the corresponding grain $x_i + \Xi_i$ (which produces the clump's boundary) either contains the grain $x_j + \Xi_j$ with $j \neq i$ or misses it. Note that the conditional probability that the boundary of a grain of radius r is not intersected by all other grains is equal to $\mathbf{P}\left\{(\partial B_r(o)) \cap \Xi = \emptyset\right\}$, since, by Slivnyak's theorem [SKM95, p. 41], all other grains have the same distribution as the original Boolean model and are independent of the chosen grain. Therefore, the density of the radius of a circular clump can be computed as

$$
\begin{aligned}
f_{cl}(r) &= f(r)\mathbf{P}\left\{(\partial B_r(o)) \cap \Xi = \emptyset\right\} \\
&= f(r)\exp\left\{-\lambda\pi\mathbf{E}\left[(r + \xi)^2 - (\max(r - \xi, 0))^2\right]\right\} \\
&= f(r)\exp\left\{-\lambda\pi\left[\mathbf{E}\xi + \int_r^\infty (x - r)^2 f(x)dx\right]\right\}, \quad r > 0.
\end{aligned}
$$

If $f(\cdot)$ is the normal density with mean μ and variance σ^2, then

$$
\int_r^\infty (x - r)^2 f(x)dx = \sigma^2 \int_s^\infty (x^2 - 2sx + s^2)\phi(x)dx = \sigma^2\psi(s),
$$

where $s = (r - \mu)/\sigma$,

$$
\psi(s) = (1 - \Phi(s))(1 + s^2) - s\phi(s),
$$

and ϕ (resp. Φ) is the density (resp. the distribution function) of the standard Gaussian random variable.

Therefore,

$$
f_{cl}(r) = \frac{1}{\sqrt{2\pi}}\exp\{-4\pi\lambda\mu^2\}\exp\left\{-\lambda\pi\left[4\mu\sigma s + \sigma^2\psi(s)\right] - \frac{s^2}{2}\right\}.
$$

Furthermore, $\psi(x)$ may be approximated by $\frac{1}{2}(x - 1)^2$, i.e.

$$
\psi(x) = \frac{1}{2}(x - 1)^2 + 0.202x + \mathcal{O}(x^3) \quad \text{as} \quad x \to 0.
$$

This approximation can be used to obtain the result that the radii of circular clumps are approximately normally distributed with mean

$$
\mu_{cl} = \frac{\mu\sigma^{-2} - \lambda\pi(3\mu - \sigma)}{\lambda\pi + \sigma^{-2}} \tag{8.3}
$$

and variance
$$\sigma_{cl}^2 = (\lambda\pi + \sigma^{-2})^{-1}. \tag{8.4}$$

This is a good approximation if $\mu > 3\sigma$ and $\lambda\pi^2\mu < 1/3$. The latter ensures a relatively small area fraction.

Furthermore, note that the area fraction satisfies

$$p = 1 - \exp\{-\lambda\pi(\mu^2 + \sigma^2)\}, \tag{8.5}$$

which is true for all distributions of the radius. Thus, (8.3), (8.4) and (8.5) give a system of 3 equations that makes it possible to estimate λ, μ and σ.

Sometimes the parametric approach allows us to estimate parameters even if lower-dimensional sections of the Boolean model are available, see [BS91] and Section 8.2.

In the anisotropic case one can assume that the grain is ξM for a deterministic set M and a scale factor ξ belonging to one of the above described parametric families. The set M is an additional *set-valued parameter*, which can be found through an estimator of the mean body of the grain, since $\mathbf{E}\Xi_0 = (\mathbf{E}\xi)M$.

It is possible also to choose one of the random polygons [SKM95, pp. 324, 328], [SS94, pp. 125–130] as the typical grain, see [QCCJ92]. These random polygons form one-parametric families of distributions. For example, the distribution of the Poisson polygon or Poisson–Dirichlet polygon depends on one parameter only (the intensity μ of the corresponding process of lines or points). Then all parameters can be estimated from two measurements of functionally independent parameters, say from the equations

$$\begin{aligned} L_A &= \lambda(1-p)\bar{U}, \\ N_A^+ &= \lambda(1-p), \end{aligned}$$

since the mean perimeter of the grain $\bar{U}$ is known for both models and is equal to $4\mu^{-1}$ or to $4\mu^{-1/2}$ for the Poisson and Poisson–Dirichlet polygons respectively.

8.2   Lower-Dimensional Sections

It is known [Mat75, p. 144], [SKM95, p. 81] that lower-dimensional sections of the Boolean model are Boolean models themselves in the corresponding lower-dimensional subspaces. From the applied point of view it is sometimes very natural to work with lower-dimensional sections: for example, in microscopy it is typical to work with planar sections of three-dimensional objects.

However, the transition to lower-dimensional sections diminishes the available information. For example, it is no longer possible to detect tangent points of a three-dimensional Boolean model by its planar sections. So-called 'thick' sections make this possible, but the corresponding theoretical approach does not differ very much from that based on the full three-dimensional observation. (It is not applicable to projected thick sections, which require the use of another more sophisticated technique, see [SKM95, p. 346].)

The difficulties begin even while estimating individual numerical parameters. Let us consider a simple example related to tangent points counting.

Let Ξ be an *isotropic* Boolean model in $\mathbf{R}^3$. Then for its planar section the following equations hold:

$$p = A_A \quad = \quad 1 - \exp\{-\lambda \bar{V}\}, \tag{8.6}$$

$$L_A \quad = \quad \lambda \pi (\bar{S}/4\bar{\bar{b}})(1 - p), \tag{8.7}$$

$$N_A^+ \quad = \quad \lambda \bar{\bar{b}}(1 - p), \tag{8.8}$$

where $\bar{V}$, $\bar{S}$ and $\bar{\bar{b}}$ are respectively the expected volume, expected surface area and expected mean width of the grain Ξ_0 in $\mathbf{R}^3$, and λ is the intensity of the three-dimensional Boolean model, see [SKM95, p. 350]. Note that $\bar{\bar{b}}$ is equal to the expected value of the width function $b(\Xi_0, u) = h(\Xi_0, u) + h(\Xi_0, -u)$ averaged for all u from the unit sphere in $\mathbf{R}^3$. In general, it is not possible to find four parameters λ, $\bar{V}$, $\bar{S}$ and $\bar{\bar{b}}$ from three equations (8.6), (8.7) and (8.8).

Similar equations are valid for linear sections of the planar Boolean model:

$$p = L_L \quad = \quad 1 - \exp\{-\lambda \bar{A}\}, \tag{8.9}$$

$$N_L^+ \quad = \quad \lambda(\bar{U}/\pi)(1 - p). \tag{8.10}$$

Here N_L^+ is the intensity of the point process of the left-end-points of segments resulting from the intersection of Ξ and a line. Again, three unknown variables must be found from two equations.

Weil [Wei93a, Wei95] derived a theoretical result that allows us to estimate the mean body of a three-dimensional Boolean model from its random planar sections. Let E be a random plane with uniformly distributed normal direction. For each realisation of E it is possible to estimate the mean body $\bar{\Xi}_{0,E}$ of the grain in the Boolean model $\Xi \cap E$ generated on the plane E. This can be done by methods described in Chapter 6. The corresponding mean body determines uniquely the mean body of the typical grain [GW92], although practical implementation of this approach is quite difficult.

Another approach uses shape assumptions on the grain. Suppose that the grain Ξ_0 is a three-dimensional ball $B_\xi(o)$ of random radius ξ with distribution function F_ξ. Then a planar section of Ξ is the planar Boolean model with a circular typical grain of radius η with distribution function F_η. These distribution functions are related by the integral equation famous from

Wicksell's problem of unfolding the distribution function of a system of three-dimensional spheres by its planar section, see [CO83, GV78, SKM95, Wei80]. This equation is

$$F_\xi(r) = 1 - \frac{2\mathbf{E}\xi}{\pi} \int\limits_{r}^{\infty} \frac{dF_\eta(x)}{\sqrt{(x^2 - r^2)}}, \quad r \geq 0,$$

see [SKM95, p. 354]. Now it is possible to estimate the function $F_\eta(x)$ or even its density by the methods of Sections 7.3 and 7.4, and, thereupon, to find the original distribution of ξ. Earlier results dealt either with non-overlapping spheres or with parametric families of distributions.

Notes to Section 8.2

General sections. Let Ξ be a Boolean model in $\mathbf{R}^d$ with convex isotropic grain Ξ_0, and let L_m be an m-dimensional affine subspace of $\mathbf{R}^d$. Then $\Xi \cap L_k$ is the Boolean model with new typical grain $\Xi_0^{(k)}$ and intensity $\lambda^{(k)}$. It follows from the general result of Matheron [Mat75, Section 5.3] that

$$\lambda^{(k)} = \lambda \frac{b_{d-k} b_k}{b_d \binom{d}{d-k}} \mathbf{E} V_{d-k}(\Xi_0),$$

and

$$\mathbf{E} V_j^{(k)}(\Xi_0^{(k)}) = \frac{\binom{k}{j}\binom{d}{d-k}}{\binom{d}{d-k+j}} \frac{b_{d-k+j}}{b_j b_{d-k}} \frac{\mathbf{E} V_{d-k+j}(\Xi_0)}{\mathbf{E} V_{d-k}(\Xi_0)},$$

where $V_j^{(k)}(\cdot)$ is the jth intrinsic volume taken in the space $\mathbf{R}^k$. Note that these formulae can be generalised for the non-Poisson case, see [Wei83].

8.3 Point-Sampling

The point-sampling scheme is the 'worst' case of lower-dimensional sections. In this case we know only whether or not $x_i \in \Xi$ for x_i belonging to a discrete set $\mathbb{X} = \{x_1, x_2, \ldots\}$ (usually $\mathbb{X}$ is a grid of points). We will call this sampling scheme 'point-sampling'.

It was noticed long ago (see [Wei80] for a survey) that it is possible to estimate the area (volume) fraction of Ξ as

$$\hat{p}_W = \frac{N(\mathbb{X} \cap \Xi \cap W)}{N(\mathbb{X} \cap W)},$$

see also Section 3.1. Unfortunately, this sampling scheme does not allow us to apply the estimation methods described above to estimate other parameters.

For example, it is not possible to detect tangent points (if the grid is sparse enough), so that all the methods based on tangent points counting are no longer applicable.

The point-sampling scheme was explored by Bortnik [Bor94]. In fact, the point-sampling approach still allows us to estimate the covariance function of the Boolean model at several points. For example, if the set $\mathbb{X}$ contains a sufficient number of points $x_i, x_i + v$, then the covariance function $C(v)$ at the point v can be estimated. Further progress is difficult without shape assumptions being imposed on the grain. Following [Bor94] suppose that the grain is the disk of radius ξ with the density $f_\xi(r)$. Then the function q from (2.18) admits the representation

$$\log q(v) = \lambda \int\limits_0^\infty k(v, r) f_\xi(r) dr . \tag{8.11}$$

The kernel $k(v, r)$ is related to the set-covariance function of the grain, which depends on $\|v\|$ only and is equal to the area of the intersection of two disks of radius r with the distance $\|v\|$ between their centres,

$$k(v, r) = \mathsf{A}(B_r(o) \cap B_r(v)) .$$

Then (8.11) can be solved to estimate the density of ξ. It should be noted that the instability of (8.11) makes it necessary to apply the so-called *stochastic regularisation* technique, see [VS78, Bor94]. A similar approach yields the density of ξ for the grain $\Xi_0 = \xi M$ with known M. Note that also in the usual (full) sampling scheme equation (8.11) is useful, since it gives the possibility of estimating the distribution of the size from the empirical covariance.

Due to its neglect of direct geometrical information the point-sampling approach is inherent to random field theory. Another random fields approach to statistics of the Boolean model was considered in [BMF86, BMF87, BM88] in relation to the study of polymer materials. The authors considered a Boolean model in $\mathbf{R}^3$ with spherical grains within the parallelepiped $[0, a] \times [0, b] \times [0, c]$ and worked with the random field $\xi(u)$, $u \in [0, a] \times [0, b]$, such that $\xi(u)$ is equal to the length of the vertical segment $\{u + (0, 0, t) : 0 \le t \le c\}$ covered by Ξ. An estimator for the radius distribution of the grain was constructed from observation of the random field $\xi(u)$.

Combinatorial properties of the Boolean model sampled on the grid are also of interest. Let us consider the isotropic Boolean model sampled on a 'regular' grid in a rectangular window W, and let M be the number of pairs of both covered bonded vertices (it is equivalently possible to use uncovered or 'half-covered' pairs of bonded vertices). Hall [Hal88, Theorem 5.1] derived an asymptotic expansion for the statistic

$$U_W = \left\{ \frac{M(n-1)}{\nu N(n-N)} - 1 \right\} \left[\frac{2(n-\nu-1)}{\nu(n-2)(n-3)} \left\{ 1 - \frac{n-1}{N(n-N)} \right\} \right]^{-1/2},$$

where ν is the degree of the lattice (the number of edges radiating from a vertex) and n (respectively N) is the total number of vertices (respectively covered vertices) inside the window W. If the minimum edge length of the regular grid is greater almost surely than the diameter of the grain, then U_W is asymptotically normal $\mathcal{N}(0,1)$ as the window W grows infinitely.

8.4 Spatial Censoring

The Boolean model Ξ can be used to build new random set models by means of set-theoretic operations, see [Ser82, pp.502–511] and [SKM95, pp. 95, 96]. For two independent Boolean models $\Xi^{(1)}$ and $\Xi^{(2)}$, their union $\Xi^{(1)} \cup \Xi^{(2)}$ is again the Boolean model, while the intersection $\Xi^{(1)} \cap \Xi^{(2)}$ and closures of the difference $(\Xi^{(1)} \setminus \Xi^{(2)})$ and the symmetric difference $((\Xi^{(1)} \setminus \Xi^{(2)}) \cup (\Xi^{(2)} \setminus \Xi^{(1)}))$ are *not* Boolean models.

These operations can be used to describe *spatial censoring* effects in statistics of the Boolean model. Then it is assumed that an observation of the set $\Xi^{(1)} \cup \Xi^{(2)}$ (or $\Xi^{(1)} \cap \Xi^{(2)}$ etc.) is available and $\Xi^{(2)}$ is a random set independent of $\Xi^{(1)}$, such that the distribution of $\Xi^{(2)}$ is known. The aim is to estimate parameters of $\Xi^{(1)}$, while the set $\Xi^{(2)}$ (in general, it can be any random closed set with a known distribution) describes the censoring effects, see [Mol92].

It is easier to work with the union-censoring, since the capacity functional $T_{\Xi^{(1)}}$ of $\Xi^{(1)}$ can be found from the capacity functionals of $\Xi^{(1)} \cup \Xi^{(2)}$ and $\Xi^{(2)}$ as

$$T_{\Xi^{(1)}}(K) = \frac{T_{\Xi^{(1)} \cup \Xi^{(2)}}(K) - T_{\Xi^{(2)}}(K)}{1 - T_{\Xi^{(2)}}(K)} \tag{8.12}$$

for all compact sets K. Thus, all estimation methods based on hitting probabilities, the covariance, and contact distribution functions work also in this case. Indeed, all terms in the right-hand side of (8.12) are either known (like $T_{\Xi^{(2)}}(K)$) or can be estimated as aggregate parameters of the union set $\Xi^{(1)} \cup \Xi^{(2)}$ (like $T_{\Xi^{(1)} \cup \Xi^{(2)}}(K)$).

If both $\Xi^{(1)}$ and $\Xi^{(2)}$ are Boolean models with convex grains, then it is possible to work out a variant of the tangent points method, since the intensity of the point process of tangent points of $\Xi^{(1)} \cup \Xi^{(2)}$ can be found as

$$(\lambda^{(1)} + \lambda^{(2)})(1 - (1 - p^{(1)})(1 - p^{(2)})),$$

where $\lambda^{(i)}$ and $p^{(i)}$ are respectively the intensity and the area fraction of $\Xi^{(i)}$, $i = 1, 2$.

Other schemes (for instance, intersection-censoring) are difficult, since it is not possible to find the capacity functional $T_{\Xi^{(1)} \cap \Xi^{(2)}}(K)$ through the capacity functionals of $\Xi^{(1)}$ and $\Xi^{(2)}$ for general K. Fortunately, this is possible for finite-point compact set K. For example, the covariance of $\Xi^{(1)}$ can be found as

$$C_{\Xi^{(1)}}(v) = C_{\Xi^{(1)} \cap \Xi^{(2)}}(v) / C_{\Xi^{(2)}}(v).$$

Therefore, statistical methods based on the covariance (see Sections 7.1, 8.3 and [Mol92]) can be applied in this framework. Again, $C_{\Xi^{(1)}}(v)$ can be found through the estimated value of $C_{\Xi^{(1)} \cap \Xi^{(2)}}(v)$ and the known value of $C_{\Xi^{(2)}}(v)$.

Such an intersection-censoring scheme arises, e.g., if inspections of $\Xi^{(1)}$ are made within $\Xi^{(2)}$ only. The set $\Xi^{(2)}$ can be viewed as the Boolean model with a circular typical grain. If its radius tends to zero, then we know only whether or not points of the Poisson point process belong to $\Xi^{(1)}$. This scheme is similar to the estimation of characteristics of a woodland region from observations at random points (e.g., birds' nests).

8.5   Multiphased Boolean Models

Suppose m independent Boolean models $\Xi^{(1)}, \ldots, \Xi^{(m)}$ to be superimposed. Each point x is then classified as being from phase 1 if it is covered by $\Xi^{(1)}$, from phase 2 if it is covered by $\Xi^{(2)}$ but not by $\Xi^{(1)}$, etc., see [Hal88, p. 295]. Then the first phase coincides with $\Xi^{(1)}$, the second is $\Xi^{(2)} \setminus \Xi^{(1)}$, etc.

This *multiphased* (or texture) model was introduced in [Ser82, pp. 503–505] and applied to the description of three-phased textures (sinter materials). In such a case one phase models the haematites, the second phase the slag, and the third the calcium ferrites, see also [SKM95, p. 96].

Such multiphased models can be considered a particular difference-censoring case, when all components with 'higher priorities' overshadow the lower ones. The covariances of all observed parts of the components can be calculated recurrently, see [Mol94a]. Thus, the covariance of $\Xi^{(1)}$ can be estimated directly, since this component is observable. Then, the covariance of $\Xi^{(2)}$ can be found from the covariances of $\Xi^{(1)}$ and $\Xi^{(2)} \setminus \Xi^{(1)}$ as

$$C_{\Xi^{(2)}}(v) = \frac{C_{\Xi^{(2)} \setminus \Xi^{(1)}}(v)}{1 - 2p^{(1)} + C_{\Xi^{(1)}}(v)},$$

and so on. Hence, covariance-based statistical methods are still applicable for multiphased Boolean models.

8.6 Discrete Boolean Models

Discrete Boolean models appear naturally when continuous Boolean models are sampled on a grid. On the other hand, the discrete framework appears when we work with computer images, which are inherently discrete. Sidiropoulos et al. [Sid92, SBB94] have studied discrete Boolean models that are subsets of the grid $\mathbb{Z}^2$ (points in the plane with integer coordinates). Some of their results are summarised below. The main difference between this discrete approach and the case of subsets of $\mathbf{R}^2$ (continuous Boolean model) is that discrete compact random sets allow us to define the probability density in a natural way. Indeed, discrete compact random set X can take no more than a countable number of values, whence the probabilities $\mathbf{P}\{X = K\}$ for all possible K serve as a density of X. In turn, this allows us to construct the maximum likelihood estimators.

We will deal with random subsets of B which is a bounded subset of the grid $\mathbb{Z}^2$ such that $o \in B$. The estimation technique is based on the so-called Möbius inversion formula [Aig79],

$$\mathbf{P}\{X = K\} = \sum_{K' \subseteq K} (-1)^{|K'|} Q_X(K^c \cup K'), \qquad (8.13)$$

where X is a random subset of B, $K \subset B$,

$$Q_X(K) = \mathbf{P}\{X \cap K = \emptyset\}$$

is the *avoiding functional* of X, $|K'|$ is the number of elements in K', and $K^c = B \setminus K$.

If Ξ_i, $i \geq 0$, is a sequence of independent identically distributed random subsets of B (grains), then

$$\Xi = \bigcup_{y_i \in \Psi} (\Xi_i + y_i)$$

is called the *discrete Boolean model* in B. Here Ψ is a binary Bernoulli random field on B, i.e. each $z \in B$ is contained in Ψ with probability $\lambda(z)$ independently of all other points. In fact, $\lambda(z)$ plays the same role as the intensity function of a non-homogeneous Poisson point process. Then

$$Q_\Xi(K) = \prod_{z \in B} \left[1 - \lambda(z) + \lambda(z) Q_{\Xi_0}(K - z) \right]. \qquad (8.14)$$

The particular case studied in [SBB94] appears if $\Xi_0 = \xi H$, where ξ is a random variable with possible values $1, \ldots, \bar{R}$ such that $\mathbf{P}\{\xi = i\} = f(i)$, $0 \leq i \leq \bar{R}$. Furthermore, H is a convex subset of B, and $\xi H = H \oplus \cdots \oplus H$. The random set Ξ is called the *discrete radial Boolean random set* with parameters

$(\lambda(\cdot), H, f(\cdot))$. This case roughly corresponds to the situation considered on p. 104 for the typical grain given by ξM for a deterministic set M.

The introduced assumptions allow us to simplify formula (8.14) for the avoiding functional $Q_\Xi(K)$. For instance, if $\lambda(z) = 1 - q$ is constant, and $\xi = 1$ almost surely, then

$$Q_\Xi(K) = q^{|K \oplus \check{H}|} .$$

Remember that $\check{H} = \{-x : x \in H\}$. Then the density given by the Möbius inversion formula (8.13) for $\Xi = X$ can be bounded as

$$L_q(K) \leq \mathbf{P}\{\Xi = K\} \leq U_q(K)$$

with

$$L_q(K) = q^{|K^c \oplus \check{H}|}(1 - q)^{|(K^c \oplus \check{H})^c|} ,$$

and

$$U_q(K) = \frac{1}{2}q^{|K^c|}\left[(1 + q)^{|K|} + (1 - q)^{|K|}\right]$$
$$- 2^{|K|-1}q^{|K^c \oplus \check{H}| + |K \oplus \check{H}|} .$$

These bounds are unimodal with the mode of the lower bound, $L_q(K)$, located at

$$\hat{q}(K) = \frac{|K^c \oplus \check{H}|}{|B|} .$$

Then $\hat{q}(K)$ can be used to estimate q from below [SBB94]. For random ξ, Sidiropoulos et al. [SBB94] argue for using the approximation

$$Q_\Xi(K) \approx q^{\mathbf{E}|K \oplus \xi \check{H}|} ,$$

which works if q is sufficiently close to 1, i.e. the grains are mostly non-overlapping and the area fraction is very low.

It should be noted that in the same situation the continuous approach from Section 8.3 based on stochastic regularisation (but still using the observations on the grid) allows estimation of the probability density of ξ. Furthermore, the continuous approach yields estimators of the mean body of the grain, i.e. the set $(\mathbf{E}\xi)H$. On the other hand the continuous approach does not allows us to formulate results in terms of the maximum likelihood estimators.

Further results on discrete (linear) Boolean models can be found in [DH95].

9

Testing the Boolean Model Assumption

Testing of the Boolean model assumption is more difficult than estimation of parameters. There are no established, strict and reliable methods that allow testing of convexity or isotropy of the grain or the Boolean model hypothesis. The existing naïve procedures are mostly based on visual comparisons and simple graphical and Monte Carlo tests. For example, it is possible to test the homogeneity by dividing the observation window into several sub-windows, estimating parameters in each window, and, finally, analysing the corresponding sample of numerical values.

From the general point of view it is even impossible to check the Boolean model hypothesis without any assumption on the grain, since the whole realisation inside the window can be considered to be a 'grain'. The methods proposed below are designed to make heuristic conclusions as to whether or not an observed image is a Boolean model with a *convex* typical grain. Further, the words 'the Boolean model hypothesis' mean 'the Boolean model with a convex typical grain hypothesis'.

9.1 Graphical Tests for Contact Distribution Functions

The most famous and widely used method is based on the polynomial expansion of contact distribution functions, see Section 5.1. For this, it is necessary to plot the empirical logarithmic spherical (quadratic) contact distribution function $\hat{H}^l_{\circ,W}(r)$ (or $\hat{H}^l_{\square,W}(r)$) and see whether or not it can be fitted well by a linear function, see [Cre91, pp. 760–774], [Hal88, pp. 292, 304–307], [Ser82, pp. 495–497] and [SKM95, pp. 86–94]. It is possible to check if the empirical contact distribution function $\hat{H}_{o,W}(r)$ can be approximated by a quadratic polynomial.

Unfortunately, computations of significance levels and confidence intervals are not possible because of complicated dependencies between values of the function $\hat{H}^l_{o,W}(r)$. The existing theoretical results [Bad80, Hei93] cannot help to design exact applicable statistical tests and find confidence regions in this situation.

It is also possible to inspect plots of the empirical logarithmic linear contact distribution function $\hat{H}^l_{u,W}(r)$ in different directions u. Under the Boolean model hypothesis, the theoretical logarithmic linear contact distribution function satisfies

$$H^l_u(r) = -\log(1 - H_u(r)) = \lambda r c_u,$$

where c_u depends on the mean difference body of the grain, see (6.2). If $r^{-1}\hat{H}^l_{u,W}(r)$ 'looks like' a constant function for any given u, then it is possible to infer that the image is the Boolean model with a convex grain, see [SKM95, pp. 89, 90]. Plotting $\hat{H}^l_{u,W}(r)$ for different u yields a graphical 'test' of the isotropy, see also [MS94b].

If independent replications of the Boolean model are available, then it is possible to construct more objective tests. First, note that under the Boolean model hypothesis the spherical contact distribution function is determined uniquely by two numerical parameters, λ and $\bar{U} = \mathbf{EU}(\Xi_0)$ (this is valid as well for other contact distribution functions $H_B(\cdot)$ with convex structuring element B if Ξ is isotropic). Thus, confidence intervals for these numerical parameters (λ and $\bar{U}$) will yield functional confidence intervals for the contact distribution functions. A variant of the method described in [MSR+94] is explained below.

It is based on the estimates $\hat{\lambda}_{W,k}$ and $\hat{\bar{U}}_{W,k}$ computed for each Boolean model $\Xi(k)$ from a series of n independent replications of Ξ in the window W using the tangent points method, see Section 5.3.

Note that the confidence region $\mathcal{D}_W$ for the pair $(\lambda, \bar{U})$ can be computed by means of a central limit theorem for the estimators of λ and $\bar{U}$, which can be proved by the method proposed in [MS94a] and [HM95]. If $\hat{\lambda}_W$ and $\hat{\bar{U}}_W$ are estimators obtained by tangent points counting, then the random vector

$$\mathsf{A}(W)^{1/2}(\hat{\lambda}_W - \lambda, \hat{\bar{U}}_W - \bar{U})$$

converges in distribution to a Gaussian random vector (ξ_1, ξ_2) with zero mean and the second moments given by

$$\mathbf{E}\xi_1^2 = \frac{\lambda}{1-p}, \tag{9.1}$$

$$\mathbf{E}\xi_2^2 = \frac{\bar{U}^2}{\lambda(1-p)} + \bar{U}^2 \int_{\mathbf{R}^2} \Big(g_{\partial\Xi}(v) - g(v)\Big) dv, \tag{9.2}$$

$$\mathbf{E}\xi_1\xi_2 \;=\; \bar{U}^2\lambda(1-p)\int_{\mathbf{R}^2} \Big(g_{\partial\Xi}(v) - g(v)\Big)dv\,, \tag{9.3}$$

where $g_{\partial\Xi}(v)$ is the pair-correlation function of the fibre process $\partial\Xi$ and $g(v)$ is the pair-correlation function of the point process of tangent points used in the estimation procedure. In fact, (9.1) has been given in Section 5.3, (9.2) was proved in [MS94a], and (9.3) can be deduced similarly to (9.1) and (9.2). Therefore, the evaluations of the confidence region for λ and $\bar{U}$ can be performed using the following scheme.

1. For each replication $\Xi(k)$, $k = 1, \dots, n$, compute estimates of numerical parameters, $\hat{\lambda}_{W,k}$, $\hat{\bar{U}}_{W,k}$, $\hat{p}_{W,k}$ and also pair-correlation functions $\hat{g}_{\partial\Xi,W,k}(v)$ and $\hat{g}_{W,k}(v)$.
2. Average the obtained estimates for $k = 1, \dots, n$. The resulting averages, $\bar{\lambda}_W$, $\bar{\bar{U}}_W$, etc., are considered to approximate theoretical values of parameters.
3. Use these averaged values instead of the theoretical values to compute the covariance matrix by (9.1)–(9.3).
4. Construct (say, spherical or elliptical) confidence region $\mathcal{D}_\xi$ for the Gaussian random vector (ξ_1, ξ_2) with the covariances given by (9.1), (9.2) and (9.3).
5. Transform $\mathcal{D}_\xi$ to the confidence region for $(\lambda, \bar{U})$ as

$$\mathcal{D}_W = \{(\lambda, \bar{U}) : \; \mathsf{A}(W)^{1/2}(\lambda - \bar{\lambda}_W, \bar{U} - \bar{\bar{U}}_W) \in \mathcal{D}_\xi\}\,.$$

The confidence region $\mathcal{D}$ yields a functional confidence region for theoretical contact distribution functions. For example, for the logarithmic spherical contact distribution function we get the confidence region

$$H_o^l(\cdot) \in \{\lambda\pi r + \lambda\bar{U} : \; (\lambda, \bar{U}) \in \mathcal{D}, \; r \geq 0\}\,.$$

Finally, if the empirical logarithmic spherical contact distribution function, $\hat{H}_{o,W,k}^l(r)$, $0 \leq r \leq r_0$, for any observation $\Xi(k)$ of Ξ does not lie within this functional confidence region, then the Boolean model hypothesis must be rejected.

Sometimes the usual testing procedures work better if the image is made coarser and 'more' discretised, see [MFR93]. This simplifies computations and makes the main features of the image clearer. This approach is justified in particularly in the study of biological specimens [MFR93, MVGF93] and is reminiscent of the multigrid approach in image analysis.

9.2   Laslett's Transform

A different approach to testing the Boolean model assumption has been suggested by Laslett, see [Cre91, pp. 765–769] and [LCL85]. This approach uses tangent points and Laslett's theorem, which states that the tangent points after a simple transformation form the homogeneous Poisson point process of the same intensity λ as the original Boolean model. For this, knowing of convexity of the grain is essential.

Let us describe Laslett's transform. Figure 9.1 shows the Boolean model Ξ inside a rectangular window W. Each lower positive tangent point x_i is projected onto the point x_i' of the left side of the rectangle. Then x_i is mapped to point x_i'' such that x_i'' belongs to the segment $[x_i', x_i]$ and the length of $[x_i', x_i'']$ is equal to the length of the uncovered part of $[x_i', x_i]$. Similarly, each point $z \in W$ is moved to the left, $z \mapsto z''$, so that $z'' \in [z', z]$ and the length of $[z', z'']$ equals the length of the covered part of $[z', z]$, where z' is the projection of z on the left side of W.

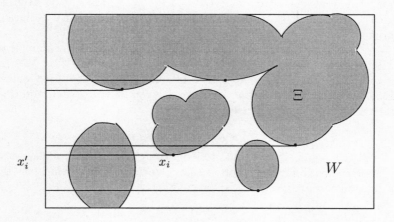

Figure 9.1 Laslett's transform applied to the tangent points.

This transformation can be applied to a vertical line in the plane (left boundary of an 'infinitely large' window) and also to the points to the left of the line (these points are moved to the right). Laslett's theorem states that the translated points form the *Poisson point process* of the intensity λ in the plane. Roughly speaking, the tangent points lose their dependencies after such a transform. This result yields the following procedure which is applicable to testing the Boolean model hypothesis.

> Take the tangent points in a certain direction u, apply to them Laslett's transform with respect to the line with direction u and test the Poisson property of the transformed point process. If it is not Poisson, then the Boolean model hypothesis must be rejected.

This procedure can also be repeated for different orientations of the window (or tangent points in different directions).

However, in practice, we have observations of the Boolean model in the *bounded* window. Since after Laslett's transform each point of the plane also moves to the left, the right-hand boundary of the window is getting distorted, see Figure 9.2. Moreover, the transformed window W' is getting *random* and *dependent* on the transformed tangent point process inside it.

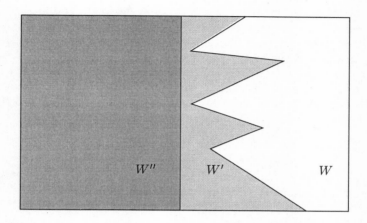

Figure 9.2   Original and transformed windows.

These complications can be partially overcome if only a rectangular part $W'' \subset W'$ is considered. However, this rectangular part is also random. For small W'', the dependency between W'' and the transformed points inside is getting weaker, but the removal of the part $W' \setminus W''$ diminishes considerably the accessible information contained in the transformed point process. Independent observations of Ξ can also help to test the Poisson hypothesis for the transformed point processes, see also [MSR+94].

9.3   Heuristic Testing of the Convexity Assumption

The first method proposed here is based on the fact that the random set $\Xi \oplus B_r(o)$ is again the Boolean model with the same intensity λ as Ξ and the typical grain $\Xi_0^r = \Xi_0 \oplus B_r(o)$. Let us note that, even for non-convex Ξ_0, its r-envelope Ξ_0^r is getting 'more convex' for larger r.

Now we can apply any estimation method based on the *convexity assumption* to the Boolean model $\Xi \oplus B_r(o)$. For the sake of definiteness let us consider the tangent points counting method and the corresponding intensity estimator. If Ξ_0 is convex, then for all sets $\Xi \oplus B_r(o)$, $r \geq 0$, the estimates of λ by this method will coincide asymptotically as the window of observations expands unboundedly. On the other hand, if Ξ_0 is not convex, then for larger r the intensity estimates will decrease with respect to r. Thus, one can plot the intensity estimates $\hat{\lambda}_W(r)$ for different $r \geq 0$ to test the convexity of the grain graphically. Moreover, the tangent points counting method allows us to construct confidence bounds for the estimated intensity based on the limit theorem for the tangent points estimator of λ, see (5.10) and (5.11).

It is also possible to perform the estimation (say, of the intensity) by two methods: one requiring the convexity assumption (e.g., tangent points method in Section 5.3) and the method of intensities in Section 5.2, which works in the more general case of simply connected grains from the convex ring. If the results are close to each other, then the convexity can be assumed. For example, if the Boolean model is isotropic, then (2.14), (3.32) and (3.10) yield

$$\mathbf{E}N_A^-(u) = \frac{L_A^2}{4\pi(1-p)}$$

under the convexity assumption. Here $N_A^-(u)$ is the intensity of the tangent points of the complement of Ξ (negative tangent points).

However, a correct implementation of this approach must include comparisons of the confidence regions based on limit theorems for estimators. Whereas such a limit theorem has been proved for the tangent points method (see Section 5.3), limiting results for the specific connectivity number and the intensity of negative tangent points are still unknown.

10

An Example

Most of the simulation studies of the methods described above have been performed in the papers where the corresponding methods were suggested. Comparative studies of intensity estimators can be found in [LS91, Sch92]. The latter also contains a survey of estimation methods. We will not perform comparative and simulation studies here, but will analyse a real picture using the statistical methods discussed above.

We will consider the sample of WC-Co alloy structure shown in Figure 2.3. This image was obtained by scanning a microphotograph provided by J.-L. Quenec'h. The scanner produced the black-and-white image in the window of 640×480 pixels on the digitised computer screen. The scanned image has been smoothed using the opening transform with the structuring element being the square of side 2. All length measurements used are in pixels. We use the fixed window $W = 640 \times 480$, and, therefore, omit the subscript W when writing up the estimates.

10.1 Area Fraction and Covariance

The area fraction was estimated by (3.1). For this, the areas are evaluated by counting the numbers of black pixels on the screen and dividing by the total number of pixels. The estimator is equal to

$$\hat{p} = 0.7503 \,.$$

The variance of the area fraction estimator was computed by (3.2) using estimator (4.13) of the covariance function. Then σ^2 from (3.2) is estimated as 410.7, whence the variance of the estimator $\hat{p}$ is

$$\operatorname{Var} \hat{p} \approx 410.7/\mathsf{A}(W) \approx 0.001337 \,,$$

and the standard deviation of $\hat{p}$ is approximately equal to 0.036.

An estimate for the function $q(\cdot)$ from (2.18) is shown in Figure 10.1. The function $\hat{q}(v)$ takes values between $(1 - \hat{p})^{-1} \approx 4.004$ (for $v = o$) and 1 (for large $\|v\|$). Note that the estimate exhibits fluctuations for larger values of v.

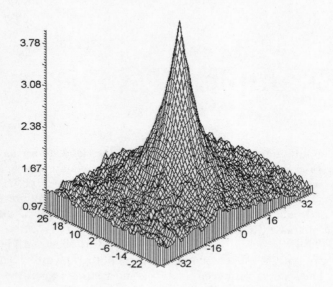

Figure 10.1   Estimate of the function $q(\cdot)$ in the window $[-40, 40] \times [-30, 30]$.

10.2   Specific Boundary Length

We estimate L_A by means of counting intersections between the fibre process $\partial\Xi$ and a system of n parallel lines with the varying direction u. The number of intersections divided by the length of the line system inside the window gives an estimator of the rose of intersections $P_L(u)$, see (3.43). Then the intensity (specific boundary length) L_A is estimated using (3.45) as the integral with respect to u. The numerical results do not depend very much on the number n. For example, $n = 100$ with 100 directions of u yields the estimate

$$\hat{L}_A = 0.05151,$$

while $n = 200$ gives $\hat{L}_A = 0.05166$ and $n = 50$ gives $\hat{L}_A = 0.05194$. We systematically omit dimension units. Clearly, L_A has the dimension of length^{-1}, in our case pixel^{-1}.

10.3 Specific Convexity and Connectivity Numbers, and Related Individual Parameters

The specific convexity and specific connectivity numbers were computed for the four main directions using (3.11) and (3.32), see Table 10.1.

Table 10.1 Specific convexity and connectivity numbers.

Direction of u	$\hat{N}_A^+(u) \times 10^{-4}$	$\hat{\chi}_A(u) \times 10^{-4}$
up	4.69	−3.19
right	4.45	−3.26
left	5.22	−3.12
down	4.02	−3.19

It is worthwhile to mention the algorithm used to detect the tangent points, since, in general, this procedure does not admit a good discrete analogue. In our case, a black pixel is considered a lower tangent point (for u directed upwards) if the surrounding pixels belong to one of the types shown in Figure 10.2.

Figure 10.2 Pixel patterns used to detect lower tangent points. The tangent point is marked by a cross.

For further estimation we use the average values for the four basic directions given in Table 10.1:

$$\hat{N}_A^+ = 4.60 \times 10^{-4} \tag{10.1}$$

and

$$\hat{\chi}_A = -3.19 \times 10^{-4}. \tag{10.2}$$

Note that estimating χ_A by counting pixel configurations on the square grid as on p. 45 gives the value $\hat{\chi}_A = -3.13 \times 10^{-4}$, which is close to the value obtained above as the difference between the numbers of positive and negative tangent points.

Before estimating functional and set-valued aggregate parameters we estimate the numerical individual parameters. The tangent points counting method for u directed upwards gives

$$\hat{\lambda}(u) = \frac{\hat{N}_A^+(u)}{1 - \hat{p}} \approx 0.00188$$

with the asymptotic variance (computed by means of (5.11))

$$\sigma_\lambda^2 \approx \frac{\hat{\lambda}(u)}{1 - \hat{p}} A(W)^{-1} \approx 2.45 \times 10^{-8}.$$

Note that we have used estimates of the parameters to compute the variance. Thus, $\sigma_\lambda \approx 1.56 \times 10^{-4}$, whence the approximate 95% confidence bound for λ based on lower tangent points counting is

$$0.00153 \leq \lambda \leq 0.00215. \tag{10.3}$$

Later we use the average tangent point estimate for the four basic directions calculated from (10.1) and given by

$$\hat{\bar{\lambda}} = 0.00184.$$

The corresponding estimates for the mean perimeter and the mean area of the grain are

$$\hat{\bar{U}} = 112.1, \quad \hat{\bar{A}} = 754.1.$$

The method of intensities based on (5.7) and (10.2) yields the estimate

$$\hat{\lambda} = 0.00211.$$

Then the mean perimeter and the mean area are estimated by

$$\hat{\bar{U}}_s = 97.7, \quad \hat{\bar{A}}_s = 657.6.$$

The intensity estimates based on counting the tangent points are usually lower than the estimates given by the method of intensities. Although a confidence interval for the latter estimate is unknown, we may conclude from (10.3) that its value agrees with the estimate obtained by counting tangent points.

Later on we will compare these estimates with the estimates obtained by the minimum contrast method. Schmitt's method does not work in this situation because of the larger grains in Figure 2.3 in comparison with the window size.

It was mentioned on p. 131 that it is possible to test the convexity of the grain heuristically by estimating the intensity for the Boolean model $\Xi \oplus B_r(o)$ for different $r \geq 0$. Table 10.2 shows that the results for $r = 0, 1, 2, 3$

Table 10.2 Estimates of λ for the Boolean model $\Xi \oplus B_r(o)$.

r	Tangent points estimate	Method of intensities
0	0.00184	0.00217
1	0.00220	0.00329
2	0.00194	0.00218
3	0.00148	0.000529

correspond to the convexity assumption. However, for larger r the estimates of intensity are definitely decreasing, which throws some doubts on the convexity assumption. This can be explained by the fact that the number of 'clumps' in the set $\Xi \oplus B_r(o)$ is getting very small, i.e. the whole set looks like one large clump with several holes.

10.4 Contact Distribution Functions and Minimum Contrast Method

The graphs of logarithmic contact distribution functions:

$$\hat{H}_o(r) = -\log(1 - \hat{H}_o(r)) \quad \text{(spherical)}$$
$$\hat{H}_\square(r) = -\log(1 - \hat{H}_\square(r)) \quad \text{(quadratic)},$$

divided by r are shown in Figures 10.3 and 10.4. The quadratic contact distribution function was computed using the square with the side length of 2 pixels as a structuring element.

The best linear regression was chosen by the method of of the least squares. The resulting estimators for the intensity and the mean perimeter and area are

$$\hat{\lambda} = 0.00183, \quad \hat{\bar{U}} = 111.7, \quad \hat{\bar{A}} = 758.2$$

(for the spherical contact distribution function), and

$$\hat{\lambda} = 0.00174, \quad \hat{\bar{U}} = 250.1, \quad \hat{\bar{A}} = 797.1$$

(for the quadratic contact distribution function).

The results obtained by using the spherical contact distribution function correspond quite well to the results by the method of intensities and by tangent points counting. The intensity estimator for the quadratic contact distribution

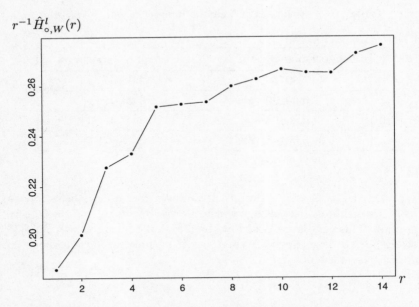

Figure 10.3   Estimate of the normalised logarithmic spherical contact
distribution function, $r^{-1} H_{\circ}^{l}(r)$.

function matches other estimates well, but the estimated expected perimeter
deviates considerably from other estimates of $\bar{U}$.

The shapes of the empirical logarithmic contact distribution functions
presented in Figures 10.3 and 10.4 do not exhibit exact linear behaviour.
This can be explained by larger statistical errors due to high area fractions of
the sets $\Xi \oplus B_r(o)$. To obtain more stable results using the minimum contrast
method one needs larger windows of observations. This has also been noticed
also in the previous simulation experiments, see [Sch92].

If empirical estimates of both spherical and quadratic contact distribution
functions are not suitable due to high area fractions, the linear contact
distribution function can help to achieve stable statistical procedures.
Figure 10.5 shows the graphs of logarithmic linear contact distribution
functions for the unit vector u directed to the right and upwards. The curves
are close to each other and almost horizontal. This speaks in favour of the
Boolean model hypothesis with isotropic grains.

Theoretically, the isotropy assumption and (5.3) yield

$$\frac{H_u^l(r)}{r} = \lambda \bar{U}/\pi = \frac{L_A}{(1-p)\pi}.$$

$r^{-1}\hat{H}^l_{\square,W}(r)$

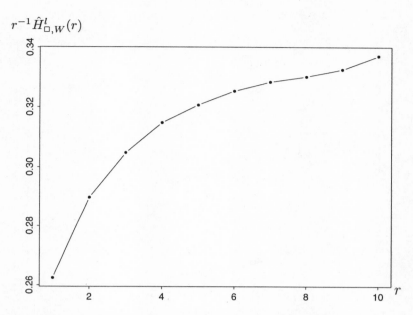

Figure 10.4 Estimate of the normalised logarithmic quadratic contact
distribution function, $r^{-1}H^l_{\square}(r)$).

This corresponds quite well to the direct estimator of L_A, since $\hat{L}_A/((1 - \hat{p})\pi) \approx 0.0657$. According to (6.2), $\hat{H}^l_u(r)/r$ gives the support function of the mean difference body. Note that the evaluation of the logarithmic linear contact distribution function for different directions u gives an estimator of the mean difference body similar to that shown in Figure 10.6b.

10.5   Set-Valued Aggregate Parameters and the Mean Difference Body

Now let us consider estimates of some set-valued parameters. We did not evaluate the convexification by the algorithm described in Section 3.5, since it is quite difficult to implement this algorithm in practice. Some reliable practical algorithms for computing the convexification need to be developed.

Figure 10.6 shows the mean star set and the Steiner compact of the fibre process $\partial\Xi$. The measurements are given in pixels.

Both look like disks, whence the mean difference body of the grain is also a disk. This argues in favour of the isotropy assumption. Note that the mean

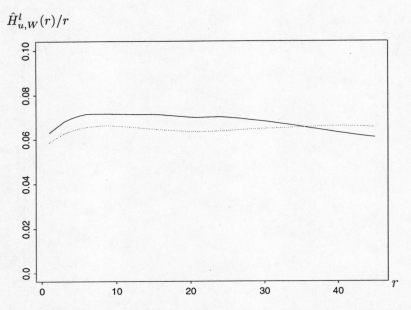

$\hat{H}_{u,W}^l(r)/r$

Figure 10.5 Estimate of the normalised logarithmic linear contact distribution functions $H_u^l(r)/r$ for two directions of u (solid line – u directed rightwards; dotted line – u directed upwards).

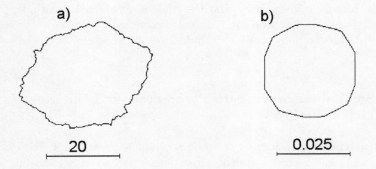

Figure 10.6 Mean star (a) and the Steiner compact (b).

difference body of the grain is equal to the rotation of the Steiner compact to $\pi/2$ with the scaling by $(\lambda(1-p))^{-1}$, see (6.5). Furthermore, the mean star has the radius-vector function inversely proportional to the support function of the mean difference body.

10.6 Pair-Correlation Function and Covering Probabilities

The pair-correlation function $g_u(v)$ of the points process $N^+(u)$ of lower positive tangent points (for u directed upwards) is estimated by (4.27) with the bandwidth ε determined from the estimated intensity $\hat{N}_A^+(u)$ of the point process $N^+(u)$ by

$$\varepsilon = 0.5/(\sqrt{\hat{N}_A^+(u)})\,.$$

This value is larger than the bandwidth recommended in [SS94, p. 285] in order to make the estimate of the pair-correlation function smoother.

The pair-correlation function is used to estimate the covering probabilities $t_u(v) = \mathbf{P}\{v \in \Xi_0^u\}$ of the shifted grain Ξ_0^u as was described in Section 7.3. Formula (7.8) yields the estimate of the covering probabilities shown in Figure 10.7. The quality of estimation of $t_u(v)$ for large $\|v\|$ is bad (even negative values may appear) because of the bad tail behaviour of both $\hat{g}_u(v)$ and $\hat{q}(v)$ used in (7.9).

10.7 Parametric Models

It is plausible to assume that the image depicted in Figure 2.3 is a Boolean model with a polygonal typical grain. Let us consider two of the simplest models of random polygons: the Poisson polygon generated by a Poisson network of lines and the Dirichlet polygon generated by the Poisson–Voronoi tessellation. Both are determined by one parameter: the intensity μ (of either the Poisson process of lines or the Poisson process of points). The expected areas and perimeters are given by

$$\bar{A} = \frac{4}{\pi\mu^2}\,, \quad \bar{U} = \frac{4}{\mu}$$

for the Poisson polygon, and

$$\bar{A} = \mu^{-1}\,, \quad \bar{U} = 4\mu^{-1/2}$$

for the Dirichlet polygon. To eliminate μ, notice that $\bar{U}^2/\bar{A} = 4\pi \approx 12.57$ for the Poisson polygon and $\bar{U}^2/\bar{A} = 16$ for the Dirichlet polygon.

Let us use the estimate for the mean area and the mean perimeter obtained above to compute the values of $\hat{\bar{U}}^2/\hat{\bar{A}}$.

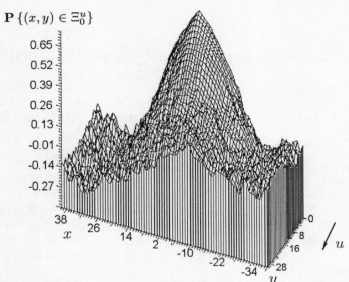

$\mathbf{P}\{(x, y) \in \Xi_0^u\}$

Figure 10.7 Covering probabilities of the shifted grain Ξ_0^u (u is directed along the ordinate axis).

Table 10.3 The characteristic parameter of the grain for different estimators.

Estimation method	$\hat{\bar{U}}^2 / \hat{\bar{A}}$
Tangent points counting	16.7
Method of intensities	14.5
Spherical contact distribution	16.45

These results argue in favour of the hypothesis that the typical grain is the Dirichlet polygon. This polygon is obtained as the typical polygon in the Poisson–Voronoi tessellation generated by the Poisson point process of intensity $\mu = (4/\bar{U})^2$. We use the estimate of expected perimeter by the tangent points method to evaluate

$$\hat{\mu} = (4/\hat{\bar{U}})^2 \approx 0.00127\,.$$

Note that Quenec'h et al. [QCCJ92] assume that the typical grain in a similar image to Figure 2.3 is the Poisson polygon. However, Table 10.3 leads us to the conclusion that the Poisson polygon hypothesis must be rejected in favour of the Dirichlet polygon hypothesis. Other models of random polyhedra (but not in relation to the Boolean model framework) are considered in [MS96].

11

Concluding Remarks

The Boolean model is the simplest model of stationary random closed sets. This is explained by strong independence assumptions: the germs have no interactions, the grains are independent of each other and of germs. In practice, these conjectures are sometimes unrealistic. In general, the Boolean model assumption is rarely fulfilled for biological objects 'in the small', since they exhibit more regular organisation. The usual fields of application of the Boolean model are the structure of materials and also natural objects 'in the large' (such as on geographical maps).

More sophisticated models use different thinning procedures and point processes with interactions, see [BM89, Han81, SKM95]. However, the corresponding statistical methods have not been developed. This situation is the same as in statistics of spatial point processes, where tractable results are obtained only under the Poisson or similar assumptions [Kar86, SKM95].

Even in the statistics of the Boolean model there are still many important open problems. Some of them are briefly described below.

1. Efficiency of estimators and lower bounds for the variances are necessary to compare different estimators theoretically. Otherwise the only way is one based on simulation studies. It is important to know what is the lowest possible variance of an unbiased intensity estimator. Also a good concept of a sufficient statistic is necessary.
2. Testing procedures for the Boolean model hypothesis are still not satisfactorily developed. Theoretical computations of significance levels are mostly needed.
3. It is necessary to prove limit theorems for the estimators obtained by the method of intensities. The main problem is related to the lack of 'second-order' integral-geometric formulae, see [Wei83].
4. As regards asymptotic properties of set-valued estimators, it is usually

possible to prove the strong consistency, but the corresponding limit theorems are not known at all. In general, it is difficult to define and compute variances of set-valued estimators.

5. Statistical estimation for special types of grains must be pursued further. The most important example is $\Xi_0 = \xi(\omega M)$, i.e. Ξ_0 is the isotropic grain that is equal to the randomly scaled and rotated deterministic set M. Then the set M itself as a set-valued parameter and the moments (or distribution) of ξ must be estimated. Also other models of random compact sets [Mol93a, SS94] can be considered.

References

[Abb78] Abbé E. (1878) Über Blutkörperzählung. *Sitzungsberichte Jenische Ges. Medizin Naturwiss.* pages 98–105.

[AFM89] Ayala G., Ferrandiz J., and Montes F. (1989) Two methods of estimation in Boolean models. *Acta Stereologica* 8(2): 629–634.

[AFM90] Ayala G., Ferrandiz J., and Montes F. (1990) Boolean models: maximum likelihood estimation from circular clumps. *Biometrical J.* 32: 73–78.

[Ahu78] Ahuja N. (1978) Geometrical properties of bombing patterns. Computer Science Technical Report TR-673, University of Maryland.

[Aig79] Aigner M. (1979) *Combinatorial Theory.* Springer, New York.

[Ald89] Aldous D. (1989) *Probability Approximations via the Poisson Clumping Heuristic.* Springer, New York.

[AM93] Archambault S. and Moore M. (1993) Statistiques morphologiques pour l'ajustement d'images. *Internat. Statist. Rev.* 61: 283–297.

[Aum65] Aumann R.J. (1965) Integrals of set-valued functions. *J. Math. Anal. Appl.* 12: 1–12.

[AV75] Artstein Z. and Vitale R.A. (1975) A strong law of large numbers for random compact sets. *Ann. Probab.* 3: 879–882.

[Aya88] Ayala G. (1988) *Inferencia en Modelos Booleanos.* PhD thesis, Universitad de València.

[Bad80] Baddeley A.J. (1980) A limit theorem for statistics of spatial data. *Adv. in Appl. Probab.* 12: 447–461.

[Bad86] Baddeley A.J. (1986) Stochastic geometry and image analysis. In *Mathematics and Computer Science, Proc. CWI Symp., 1983*, volume 1, pages 1–18. Centrum voor Wiskunde en Informatica, Amsterdam.

[BCO94] Beneš V., Chadœuf J., and Ohser J. (1994) On some characteristics of anisotropic fibre processes. *Math. Nachr.* 169: 5–17.

[Ber93] Berkhout R.J. (November 1993) An Adjustment of Haldorsen's Object-Based Conditional Simulation Algorithm. MSc Thesis, Faculty of Mining Engineering and Petroleum Exploration, Delft University of Technology.

[Bes78] Besag J. (1978) Some methods of statistical analysis for spatial data. *Bull. Inst. Intern. Statist.* 2: 77–92.

[Bes86] Besag J. (1986) On the statistical analysis of dirty pictures (with discussion). *J. Roy. Statist. Soc. Ser. B* 48: 259–302.

[BG93] Baddeley A.J. and Gill R.D. (1993) Kaplan–Meier estimators of interpoint distance distributions for spatial point processes. Technical Report BS-R9315, Centrum voor Wiskunde en Informatica, Amsterdam.

[BG94] Baddeley A.J. and Gill R.D. (1994) The empty space hazard of a spatial pattern. Technical Report 1994/3, The University of Western Australia,

Department of Mathematics.

[Bil68] Billingsley P. (1968) *Convergence of Probability Measures*. Wiley, New York.

[BM88] Bulinskaya E.V. and Molchanov S.A. (1988) Evaluation of parameters of complex random sets. III. *Moscow University Math. Bulletin* 43(6): 29–33.

[BM89] Baddeley A.J. and Møller J. (1989) Nearest-neighbour Markov point processes and random sets. *Internat. Statist. Rev.* 57: 89–121.

[BMF86] Bulinskaya E.V., Molchanov S.A., and Feraig N. (1986) Parameters estimates for complex random sets. I. *Moscow University Math. Bulletin* 41(4): 16–21.

[BMF87] Bulinskaya E.V., Molchanov S.A., and Feraig N. (1987) On an estimate of the parameters of complex random sets. II. *Moscow University Math. Bulletin* 42(2): 6–11.

[Bor94] Bortnik D. (1994) Zur Parameterschätzung in Booleschen Modellen. Diplomarbeit, Universität Kiel.

[Bri75] Brillinger D.R. (1975) Statistical inferences for stationary point processes. In Puri M. (ed) *Stochastic Processes and Related Topics*, pages 55–59. Academic Press, New York.

[BS91] Bindrich U. and Stoyan D. (1991) Stereology for pores in white bread: statistical analyses for the Boolean model by serial sections. *J. Microscopy* 162: 231–239.

[BvL95] Baddeley A.J. and Van Lieshout M.N.M. (1995) Area-interaction point processes. *Ann. Inst. Statist. Math.* 47: 601–619.

[CB94] Chadœuf J. and Beneš V. (1994) On some estimation variances in spatial statistics. *Kybernetika* 30: 245–262.

[CCH95] Cowan R., Chiu S.N., and Holst L. (1995) A limit theorem for the replication time of a DNA molecule. *J. Appl. Probab.* 32: 296–303.

[CH92] Cressie N.A.C. and Hulting F.L. (1992) A spatial statistical analysis of tumor growth. *J. Amer. Statist. Assoc.* 87: 272–283.

[Che95] Chessa A.G. (1995) *Conditional Simulation of Spatial Stochastic Models for Reservoir Heterogeneity*. PhD thesis, Delft University of Technology.

[Chi95] Chiu S.N. (1995) Limit theorems for the time of completion of Johnson–Mehl tessellations. *Adv. in Appl. Probab.* 27: 889–910.

[CI80] Cox D.R. and Isham V. (1980) *Point Processes*. Chapman and Hall, London.

[CJ94] Cabo A.J. and Janssen R.H.P. (July 1994) Cross-covariance functions characterise bounded regular closed sets. Report BS-R9426, Centrum voor Wiskunde en Informatica, Amsterdam.

[CL87] Cressie N.A.C. and Laslett G.M. (1987) Random set theory and problems of modeling. *SIAM Rev.* 29: 557–574.

[CO83] Cruz-Orive L.M. (1983) Distribution-free estimation of sphere size distribution from slabs showing overprojections and truncation, with a review of previous methods. *J. Microscopy* 131: 265–290.

[Cre79] Cressie N.A.C. (1979) A central limit theorem for random sets. *Z. Wahrsch. verw. Gebiete* 49: 37–47.

[Cre91] Cressie N.A.C. (1991) *Statistics for Spatial Data*. Wiley, New York.

[CS97] Chiu S.N. and Stoyan D. (1997) Estimators of distance distributions for spatial patterns. Statistica Neerlandica.

[CV77] Castaing C. and Valadier M. (1977) *Convex Analysis and Measurable Multifunctions*, volume 580 of *Lect. Notes Math.* Springer, Berlin.

[Dav76] Davy P.J. (1976) Projected thick sections through multi-dimensional particle aggregates. *J. Appl. Probab.* 13: 714–722.

[Dav78] Davy P.J. (1978) Aspects of random set theory. *Adv. in Appl. Probab.*

Supplement volume 10: 28–35.

[DH73] Duda R.O. and Hart P.E. (1973) *Pattern Classification and Scene Analysis.* Wiley, New York.

[DH95] Dougherty E.R. and Handley J.C. (1995) Recursive maximum-likelihood estimation in one-dimensional discrete Boolean random set model. *Signal Processing* 43: 1–15.

[Dig81] Diggle P.J. (1981) Binary mosaics and the spatial pattern of heather. *Biometrics* 37: 531–539.

[Dig83] Diggle P.J. (1983) *Statistical Analysis of Spatial Point Patterns.* Academic Press, London.

[DM77] Davy P.J. and Miles R.E. (1977) Sampling theory for opaque spatial specimens. *J. Roy. Statist. Soc. Ser.* B 39: 59–65.

[Dod71] Dodson C.T.J. (1971) Spatial variability and the theory of sampling in random fibrous networks. *J. Roy. Statist. Soc. Ser.* B 33: 88–94.

[Dud84] Dudley R.M. (1984) A course on empirical processes. In *École d'Été de Probabilités de Saint Flour XII*, volume 1097 of *Lect. Notes Math.*, pages 1–142. Springer, Berlin.

[Dup80] Dupač V. (1980) Parameter extimation in the Poisson field of discs. *Biometrika* 67: 187–190.

[DVJ88] Daley D.J. and Vere-Jones D. (1988) *An Introduction to the Theory of Point Processes.* Springer, New York.

[Eva90] Evans S.N. (1990) Rescaling the vacancy of a Boolean coverage process. In *Semin. Stoch. Proc., San Diego Calif., March 30 – April 1, 1989*, pages 23–33.

[Fuk72] Fukunaga K. (1972) *Introduction to Statistical Pattern Recognition.* Academic Press, New York.

[Gal87] Galway L.A. (1987) *Statistical Analysis of Star-Shaped Sets.* PhD thesis, Carnegie-Mellon University, Pittsburgh (PA), U.S.A.

[Gey93] Geyer C.J. (1993) Practical markov chain monte carlo (with discussion). *Statist. Sci.* 8: 473–483.

[GG84] Geman S. and Geman D. (1984) Stochastic relaxation, Gibbs distributions, and the Bayesian restoration of images. *IEEE Trans. Pattern Analysis and Machine Intelligence* 6: 721–741.

[GHZ83] Giné E., Hahn M.G., and Zinn J. (1983) Limit theorems for random sets: application of probability in Banach space results. *Lect. Notes Math.* 990: 112–135.

[Gil62] Gilbert E.N. (1962) Random subdivisions of space into crystals. *Ann. Math. Statist.* 33: 958–972.

[Gil95] Gille W. (1995) Diameter distribution of spherical primary grains in the Boolean model from small-angle scattering. *Part. Part. Syst. Charact.* 12: 123–131.

[Gir82] Girling A.J. (1982) Approximate variances associated with random configurations of hard spheres. *J. Appl. Probab.* 19: 588–596.

[GJ90] Gill R.D. and Johansen S. (1990) A survey of product-integration with a view toward application in survival analysis. *Ann. Statist.* 18: 1501–1555.

[God91] Godtliebsen F. (1991) Noise reduction using Markov random fields. *J. Magnetic Resonance* 92: 102–114.

[Gro78] Groemer H. (1978) On the extension of additive functionals on classes of convex sets. *Pacific J. Math.* 75: 397–410.

[GRS96] Gilks W.R., Richardson S., and Spiegelhalter D.J. (eds) (1996) *Markov Chain Monte Carlo in Practice.* Chapman and Hall, London.

[GV78] Goldman A. and Visscher W. (1978) Applications of integral equations in particle-size statistics. In Goldberg M.A. (ed) *Solution Methods for Integral*

Equations, pages 169–182. Plenum Press, New York/London.

[GW92] Goodey P. and Weil W. (1992) The determination of convex bodies from the mean of random sections. *Math. Proc. Cambridge Philos. Soc.* 112: 419–430.

[GW93a] Goodey P. and Weil W. (1993) Zonoids and generalizations. In Gruber and Wills [GW93b], pages 1299–1326.

[GW93b] Gruber P.M. and Wills J.M. (eds) (1993) *Handbook of Convex Geometry. Vol. A, B.* North-Holland, Amsterdam.

[Hal85a] Hall P. (1985) Distribution of size, structure and number of vacant region in a high-intensity mosaic. *Z. Wahrsch. verw. Gebiete* 70: 237–261.

[Hal85b] Hall P. (1985) On continuum percolation. *Ann. Probab.* 13: 1250–1266.

[Hal85c] Hall P. (1985) On the coverage of k-dimensional space by k-dimensional spheres. *Ann. Probab.* 13: 991–1002.

[Hal88] Hall P. (1988) *Introduction to the Theory of Coverage Processes.* Wiley, New York.

[Han81] Hanisch K.-H. (1981) On classes of random sets and point processes. *Serdica* 7: 160–166.

[Han85] Hanisch K.-H. (1985) On the second-order analysis of stationary and isotropic planar fibre processes by a line intercept method. In Nagel [Nag85], pages 141–146.

[Han95] Hansen M. (1995) *Spatial Statistics and the Variation of the Protein Network during Milk Processing.* PhD thesis, Royal Veterinary and Agricultural University, Copenhagen. Dina Research Report No. 30.

[Hei88a] Heinrich L. (1988) Asymptotic behaviour of an empirical nearest-neighbour distance function for stationary Poisson cluster process. *Math. Nachr.* 136: 131–148.

[Hei88b] Heinrich L. (1988) Asymptotic Gaussianity of some estimators for reduced factorial moment measures and product densities of stationary Poisson cluster processes. *Statistics* 19: 87–106.

[Hei92a] Heijmans H.J.A.M. (1992) Discretization of morphological operators. *J. Visual Communication and Image Representation* 3: 182–193.

[Hei92b] Heinrich L. (1992) On existence and mixing properties of germ–grain models. *Statistics* 23: 271–286.

[Hei93] Heinrich L. (1993) Asymptotic properties of minimum contrast estimators for parameters of Boolean models. *Metrika* 31: 349–360.

[Hei94a] Heijmans H.J.A.M. (1994) *Morphological Image Operators.* Academic Press, Boston.

[Hei94b] Heinrich L. (1994) Normal approximation for some mean-value estimates of absolutely regular tessellations. *Math. Methods of Statist.* 3: 1–24.

[Hei95] Heijmans H.J.A.M. (1995) Mathematic morphology: a modern approach in image processing based on algebra and geometry. *SIAM Rev.* 37: 1–36.

[HG88] Hüsler J. and Glauser K. (1988) On the coverage problem in higher dimensions with non-uniform density. *J. Microscopy* 151: 257–264.

[HGB96] Hansen M.B., Gill R.D., and Baddeley A.J. (1996) Kaplan–Meier type estimators for linear contact distributions. *Scand. J. Statist.* 23: 129–155.

[HM95] Heinrich L. and Molchanov I.S. (June 1995) Central limit theorem for a class of random measures associated with germ–grain models. Report BS-R9518, Centrum voor Wiskunde en Informatica, Amsterdam.

[HNRW89] Hübler A., Nagel W., Ripley B., and Werner G. (eds) (1989) *Geobild '89*, volume 51 of *Math. Res.*, Berlin, Akademie-Verlag.

[Hüs82] Hüsler J. (1982) Random coverage of the circle and asymptotic distributions. *J. Appl. Probab.* 19: 578–587.

[Jen93] Jensen J.L. (1993) Asymptotic normality of estimates in spatial point processes. *Scand. J. Statist.* 20: 97–109.

[JM91] Jensen J.L. and Møller J. (1991) Pseudolikelihood for exponential family models of spatial processes. *Ann. Appl. Probab.* 1: 445–461.

[Jol80] Jolivet E. (1980) Central limit theorem and convergence of empirical processes for stationary point processes. In Bartfai P. and Tomko J. (eds) *Point Processes and Queuing Problems*, pages 117–161. North-Holland, Amsterdam.

[Jol84] Jolivet E. (1984) Upper bounds of the speed of convergence of moment density estimators for stationary point processes. *Metrika* 31: 349–360.

[Jol86] Jolivet E. (1986) Parametric estimation of the covariance density for a stationary point process on $\mathbf{R}^d$. *Stochastic Process. Appl.* 22: 111–119.

[Jol91] Jolivet E. (1991) Moment estimation for stationary point processes in $\mathbf{R}^d$. In Possolo [Pos91], pages 138–149.

[Kar86] Karr A.F. (1986) *Point Processes and Their Statistical Inference.* Marcel Dekker, New York.

[Kel83] Kellerer A.M. (1983) On the number of clumps resulting from the overlap of randomly placed figures in a plane. *J. Appl. Probab.* 22: 68–81.

[Kel84] Kellerer H.G. (1984) Minkowski functionals of Poisson processes. *Z. Wahrsch. verw. Gebiete* 67: 63–84.

[Kel85] Kellerer A.M. (1985) Counting figures in planar random configurations. *J. Appl. Probab.* 22: 68–81.

[KM63] Kendall M.G. and Moran P.A.P. (1963) *Geometrical Probability.* Charles Griffin, London.

[Kol37] Kolmogorov A.N. (1937) On statistical theory of metal crystallization. *Izvestia Academy of Science, USSR, Ser. Math.* 3: 355–360. In Russian.

[Kri82] Krickeberg K. (1982) Processus ponctuels en statistique. In Hennequin P.L. (ed) *École d'Été de Probabiliteés de Saint-Flour X-1980*, volume 929 of *Lect. Notes Math.* Springer, Berlin.

[KS91] König D. and Schmidt V. (1991) *Zufällige Punktprozesse.* Teubner, Stuttgart.

[LCL85] Laslett G.M., Cressie N., and Liow S. (1985) Intensity estimation in a spatial model of overlapping particles. Unpublished manuscript, Division of Mathematics and Statistics, CSIRO, Melbourne.

[Lei80] Leichtweiss K. (1980) *Konvexe Mengen.* VEB Deutscher Verlag der Wissenschaften, Berlin.

[LL85] Liesek B. and Liesek M. (1985) A new method for testing whether a point process is Poisson. *Statistics* 16: 445–450.

[Loè63] Loève M. (1963) *Probability Theory.* Van Nostrand, Princeton, NJ, 3rd edition.

[LR90] Lešanovský A. and Rataj J. (1990) Determination of compact sets in Euclidean space by the volume of their dilation. In *Proceedings DIANA III*, pages 165–177. Mathematical Institute of ČSAV, Benchyně, Czechoslovakia.

[LS91] Lantuejoul C. and Schmitt M. (1991) Use of two new formulae to estimate the Poisson intensity of a Boolean model. In *Treizième Colloque GRETSI – Juan-Les-Pins du 16 au 20 Septembre 1991*, pages 1045–1048.

[Lya82] Lyashenko N.N. (1982) Limit theorems for sum of independent compact random subsets. *J. Soviet Math.* 20: 2187–2196.

[Lya86] Lyashenko N.N. (1986) Random 'grainy' patterns. I. *Zap. Nauchn. Sem. Leningrad. Otdel. Mat. Inst. Steklov. (LOMI)* 153: 73–96.

[Lya88] Lyashenko N.N. (1988) Random 'grainy' patterns. II. *Zap. Nauchn. Sem. Leningrad. Otdel. Mat. Inst. Steklov. (LOMI)* 166: 91–102. In Russian.

[MA91] Moore M. and Archambault S. (1991) On the asymptotic behavior and some statistics based on morphological operations. In Possolo [Pos91], pages 258–274.

[Mas82] Mase S. (1982) Asymptotic properties of stereological estimators of volume fraction for stationary random sets. *J. Appl. Probab.* 19: 111–126.

[Mat75] Matheron G. (1975) *Random Sets and Integral Geometry.* Wiley, New York.

[Mat86] Matheron G. (February 1986) Le covariogramme géométrique des compacts convexes de $\mathbf{R}^2$. Technical Report 2/86, Ecole des Mines de Paris, Centre de Géostatistique.

[MC95] Micheletti A. and Capasso V. (1995) The stochastic geometry of polymer crystallization process. Technical Report Quaderno n. 20, Università Degli Studi di Milano, Dipartamento di Mathematica, Milan.

[MFR93] Mattfeldt T., Frey H., and Rose C. (1993) Second-order stereology of benign and malignant alterations of the human mammary gland. *J. Microscopy* 171: 143–151.

[Mil69] Miles R. (1969) The asymptotic values of certain coverage probabilities. *Biometrika* 56: 661–680.

[Mil76] Miles R. (1976) Estimating aggregate and overall characterictics from thick sections by transmission microscopy. *J. Microscopy* 107: 227–233.

[MOK95] Molchanov I.S., Omey E., and Kozarovitzky E. (1995) An elementary renewal theorem for random convex compact sets. *Adv. in Appl. Probab.* 27: 931–942.

[Mol87] Molchanov I.S. (1987) Uniform laws of large numbers for empirical associated functionals of random closed sets. *Theory Probab. Appl.* 32: 556–559.

[Mol89] Molchanov I.S. (1989) On convergence of empirical accompanying functionals of stationary random sets. *Theory Probab. Math. Statist.* 38: 107–109. Translation from *Teor. Veroyatn. Mat. Stat.* 38: 97–99, 1988.

[Mol90a] Molchanov I.S. (1990) A characterization of the universal classes in the Glivenko–Cantelli theorem for random closed sets. *Theory Probab. Math. Statist.* 41: 85–89. Translation from *Teor. Veroyatn. Mat. Stat.* 41: 74–78, 1989.

[Mol90b] Molchanov I.S. (1990) Empirical estimation of distribution quantiles of random closed sets. *Theory Probab. Appl.* 35: 594–600.

[Mol90c] Molchanov I.S. (1990) Estimation of the size distribution of spherical grains in the Boolean model. *Biometrical J.* 32: 877–886.

[Mol91a] Molchanov I.S. (1991) A consistent estimate of parameters of Boolean models of random closed sets. *Theory Probab. Appl.* 36: 600–607.

[Mol91b] Molchanov I.S. (1991) Estimation of the parameters of superpositions of Poisson point processes. *Theory Probab. Math. Statist.* 43: 123–130. Translation from *Teor. Veroyatn. Mat. Stat.* 43: 111–117, 1990.

[Mol91c] Molchanov I.S. (1991) Random sets. A survey of results and applications. *Ukrainian Math. J.* 43: 1477–1487.

[Mol92] Molchanov I.S. (1992) Handling with spatial censored observations in statistics of Boolean models of random sets. *Biometrical J.* 34: 617–631.

[Mol93a] Molchanov I.S. (1993) *Limit Theorems for Unions of Random Closed Sets*, volume 1561 of *Lect. Notes Math.* Springer, Berlin.

[Mol93b] Molchanov I.S. (1993) On estimation of the shape of an isotropic grain in the Boolean model. *Acta Stereologica* 12(2, Part 1): 181–184.

[Mol94a] Molchanov I.S. (1994) On statistical analysis of Boolean models with non-random grains. *Scand. J. Statist.* 21: 73–82.

[Mol94b] Molchanov I.S. (1994) Statistical analysis of Boolean models with a randomly scaled grain. *Theory Probab. Math. Statist.* 48: 105–110.

[Mol95] Molchanov I.S. (1995) Statistics of the Boolean model: from the estimation of means to the estimation of distributions. *Adv. in Appl. Probab.* 27: 63–86.

[Mol96a] Molchanov I.S. (1996) A limit theorem for scaled vacancies of the Boolean model. *Stochastics, Stochastic Reports.*

[Mol96b] Molchanov I.S. (1996) Set-valued estimators for mean bodies related to Boolean models. *Statistics* 28: 43–56.

[MR94] Meester R. and Roy R. (1994) Uniqueness of unbounded occupied and vacant components in Boolean models. *Ann. Appl. Probab.* 4: 933–951.

[MS80] Mecke J. and Stoyan D. (1980) Formulas for stationary planar fibre processes I – General theory. *Math. Operationsforsch. Statist. Ser. Statistik* 12: 267–279.

[MS94a] Molchanov I.S. and Stoyan D. (1994) Asymptotic properties of estimators for parameters of the Boolean model. *Adv. in Appl. Probab.* 26: 301–323.

[MS94b] Molchanov I.S. and Stoyan D. (1994) Directional analysis of fibre processes related to Boolean models. *Metrika* 41: 183–199.

[MS96] Molchanov I.S. and Stoyan D. (1996) Statistical models of random polyhedra. *Comm. Statist. Stochastic Models* 12: 199–214.

[MSF93] Molchanov I.S., Stoyan D., and Fyodorov K.M. (1993) Directional analysis of planar fibre networks: Application to cardboard microstructure. *J. Microscopy* 172: 257–261.

[MSR+94] Mattfeldt T., Schmidt V., Reepschläger D., Rose C., and Frey H. (1994) Second-order stereology of benign and malignant glandular tissue: estimation of stochastic-geometric functions and tests for the Boolean model. Unpublished manuscript.

[MVGF93] Mattfeldt T., Vogel U., Gottfried H.-W., and Frey H. (1993) Second-order stereology of prostatic adenocarcinoma and normal prostatic tissue. *Acta Stereologica* 12: 203–208.

[Nag85] Nagel W. (ed) (1985) *Geobild '85, Workshop on Geometrical Problems in Imaging, Georgenthal (GDR), January, 14–18, 1985*, Wissenschaftliche Beiträge der Friedrich-Schiller-Universität Jena, Jena.

[Nag93] Nagel W. (1993) Orientation-dependent chord length distributions characterize convex polygons. *J. Appl. Probab.* 30: 730–736.

[Nau79] Naus J.L. (1979) An indexed bibliography of clusters, clumps and coincidences. *Internat. Statist. Rev.* 47: 47–78.

[NS72] Neyman J. and Scott E.L. (1972) Processes of clustering and applications. In Lewis P.A.W. (ed) *Stochastic Point Processes. Statistical Analysis, Theory and Applications*, pages 646–681. Wiley, Chichester.

[NZ76] Nguyen X.X. and Zessin H. (1976) Punktprozesse mit Wechselwirkung. *Z. Wahrsch. verw. Gebiete* 37: 91–126.

[NZ79] Nguyen X.X. and Zessin H. (1979) Ergodic theorems for spatial processes. *Z. Wahrsch. verw. Gebiete* 48: 133–158.

[Oga88] Ogata Y. (1988) Statistical methods for earthquake occurrences and residual analysis for point processes. *J. Amer. Statist. Assoc.* 83: 9–27.

[Ohs83] Ohser J. (1983) On estimators for the reduced second moment measure of point processes. *Math. Operationsforsch. Statist. Ser. Statistik* 14: 63–71.

[OS80] Ohser J. and Stoyan D. (1980) Zur Beschreibung gewisser zufälliger Muster in der Geologie. *Z. angew. Geol.* 26: 209–212.

[OS81] Ohser J. and Stoyan D. (1981) On the second-order and orientation analysis of planar stationary point processes. *Biometrical J.* 23: 523–533.

[OT88] Ohser J. and Tscherny H. (1988) *Grundlagen der quantitativen Gefügeanalyse*, volume 264 of *Freiberger Forschungshefte, Reihe B*. Deutscher Verlag für Grundstoffindustrie, Leipzig.

[Pen91] Penrose M.D. (1991) On a continuum percolation model. *Adv. in Appl. Probab.* 23: 536–556.

[Pie77] Pielou E.C. (1977) *Mathematical Ecology.* Wiley, New York.

[Pos91] Possolo A. (ed) (1991) *Spatial Statistics and Imaging,* volume 20 of *IMS Lecture Notes – Monographs.* Proceedings of the Joint IMS–AMS–SIAM Summer Research Conference on Spatial Statistics and Imaging, Brunswick, Maine.

[Pra77] Pratt W.K. (1977) *Digital Image Processing.* Wiley, New York.

[PS87] Prêteux F. and Schmitt M. (May 1987) Analyse et synthese de fonctions booleennes. Théorèmes de caractérisation et démonstrations. Technical report, Ecole de Mines, Fontainebleau.

[PS88] Prêteux F. and Schmitt M. (1988) Boolean texture analysis and synthesis. In Serra J. (ed) *Image Analysis and Mathematical Morphology, Volume 2: Theoretical Advances,* pages 377–400. Academic Press, New York.

[Pyk83] Pyke R. (1983) A uniform central limit theorem for partial sum processes indexed by sets. *London Math. Soc., Lect. Notes Ser.* 79: 219–240.

[QCCJ92] Quenec'h J.-L., Coster M., Chermant J.-L., and Jeulin D. (1992) Study of the liquid-phase sintering process by probabilistic models: application to the coarsening of WC-Co cermets. *J. Microscopy* 168: 3–14.

[Rat93] Rataj J. (1993) Random distances and edge correction. *Statistics* 24: 377–385.

[Rat95] Rataj J. (1995) Characterization of compact sets by their dilation volume. *Math. Nachr.* 173: 287–295.

[Rat96] Rataj J. (1996) Estimation of oriented direction distribution of a planar body. *Adv. in Appl. Probab.* 28: 394–404.

[Rip76] Ripley B.D. (1976) The second-order analysis of stationary point processes. *J. Appl. Probab.* 13: 255–266.

[Rip86] Ripley B.D. (1986) Statistics, images and pattern recognition. *Canad. J. Statist.* 14: 83–102.

[Rob44] Robbins H.E. (1944) On the measure of a random set. I. *Ann. Math. Statist.* 15: 70–74.

[Rob45] Robbins H.E. (1945) On the measure of a random set. II. *Ann. Math. Statist.* 16: 342–347.

[RS89] Rataj J. and Saxl I. (1989) Analysis of planar anisotropy by means of the Steiner compact. *J. Appl. Probab.* 26: 490–502.

[RS92] Rataj J. and Saxl I. (1992) Estimation of direction distribution of a planar fibre system. *Acta Stereologica* 11: 631–636. Suppl. I.

[San76] Santaló L. (1976) *Integral Geometry and Geometric Probability.* Addison-Welsey, Reading, Mass.

[SBB94] Sidiropoulos N.D., Baras J.S., and Berenstein C.A. (1994) Algebraic analysis of the generating functional for discrete random sets and statistical inference for intensity in the discrete Boolean random-set model. *Journal of Mathematical Imaging and Vision* 4: 273–290.

[Sch91] Schmitt M. (1991) Estimation of the density in a stationary Boolean model. *J. Appl. Probab.* 28: 702–708.

[Sch92] Schröder M. (1992) Schätzer für Boolesche Modelle im $\mathbf{R}^2$ und $\mathbf{R}^3$. Diplomarbeit, Universität Karlsruhe.

[Sch93a] Schmitt M. (1993) On two inverse problems in mathematical morphology. In Dougherty E.R. (ed) *Mathematical Morphology in Image Processing,* volume 34 of *Opt. Engrg.,* pages 151–169. Marcel Dekker, New York.

[Sch93b] Schneider R. (1993) *Convex Bodies. The Brunn–Minkowski Theory.* Cambridge University Press, Cambridge.

[Ser82] Serra J. (1982) *Image Analysis and Mathematical Morphology.* Academic Press, London.

[SH82] Siegel A.F. and Holst L. (1982) Covering the circle with random arcs of random sizes. *J. Appl. Probab.* 19: 373–381.

[She72] Shepp L.A. (1972) Covering the line with random intervals. *Z. Wahrsch. verw. Gebiete* 23: 163–170.

[Sid92] Sidiropoulos N.D. (1992) *Statistical Inference, Filtering, and Modeling of Discrete Random Sets.* PhD thesis, The University of Maryland.

[SKM95] Stoyan D., Kendall W.S., and Mecke J. (1995) *Stochastic Geometry and Its Applications.* Wiley, Chichester, second edition.

[SM97] Stoyan D. and Molchanov I.S. (1997) Set-valued means of random particles. *Journal of Mathematical Imaging and Vision* .

[SMP80] Stoyan D., Mecke J., and Pohlmann S. (1980) Formulas for stationary planar fibre processes II – partially oriented fibre systems. *Math. Operationsforsch. Statist. Ser. Statistik* 11: 281–286.

[SOS87] Schwandtke A., Ohser J., and Stoyan D. (1987) Improved estimation in planar sampling. *Acta Stereologica* 6(2): 325–334.

[SPRB94] Saxl I., Pelikan K., Rataj J., and Besterci M. (1994) *Quantification and Modelling of Heterogeneous Systems.* Cambridge Interscience Publishing, Cambridge.

[SS94] Stoyan D. and Stoyan H. (1994) *Fractals, Random Shapes and Point Fields.* Wiley, Chichester.

[Sta89] Stadje W. (1989) Coverage problems for random intervals. *SIAM J. Appl. Math.* 49(5): 1538–1551.

[Sto79] Stoyan D. (1979) Applied stochastic geometry: a survey. *Biometrical J.* 21: 693–715.

[Sto81] Stoyan D. (1981) On the second-order analysis of stationary planar fiber processes. *Math. Nachr.* 102: 183–199.

[Sto83] Stoyan D. (1983) Inequalities and bounds for variances of point processes and fibre processes. *Math. Operationsforsch. Statist. Ser. Statistik* 14: 409–419.

[Sto85] Stoyan D. (1985) Practicable methods for the determination of the pair-correlation function of fibre processes. In Nagel [Nag85], pages 131–140.

[Sto89] Stoyan D. (1989) On means, medians and variances of random sets. In Hübler et al. [HNRW89], pages 99–104.

[Sto90] Stoyan D. (1990) Stereology and stochastic geometry. *Internat. Statist. Rev.* 58: 227–242.

[Str70] Streit F. (1970) On multiple geometric integrals and their applications to probability theory. *Canad. J. Math.* 22: 151–163.

[SW83] Schneider R. and Weil W. (1983) Zonoids and related topics. In Gruber P.M. and Wills J.M. (eds) *Convexity and its Applications*, pages 296–317. Birkhäuser, Basel.

[SW86] Salinetti G. and Wets R.J.-B. (1986) On the convergence in distribution of measurable multifunctions (random sets), normal integrands, stochastic processes and stochastic infima. *Math. Oper. Res.* 11: 385–419.

[SW92] Schneider R. and Weil W. (1992) *Integralgeometrie.* B.G. Teubner, Stuttgart.

[SW93] Schneider R. and Wieacker J.A. (1993) Integral geometry. In Gruber and Wills [GW93b], pages 1349–1390.

[Ter94] Terwogt J.H. (September 1994) A 3-D conditional simulation algorithm for fluvial channel deposits. TNO-report OS 94-66-A, TNO Institute of Applied Geoscience, Delft.

[UF85] Upton G. and Fingleton B. (1985) *Spatial Data Analysis by Example.* Wiley, Chichester.

[Vit83] Vitale R.A. (1983) Some developments in the theory of random sets. *Bull. Inst. Intern. Statist.* 50: 863–871.

[Vit88] Vitale R.A. (1988) An alternate formulation of mean value for random geometric figures. *J. Microscopy* 151: 197–204.

[Vit91] Vitale R.A. (1991) The translative expectation of a random set. *J. Math. Anal. Appl.* 160: 556–562.

[VS78] Vapnik V.N. and Stefanyuk A.R. (1978) Nonparametric methods for reconstructing probability densities. *Automat. Remote Control* 39: 1127–1140. Translation from *Avtomatika Telemekh.*, 1978, No. 8, 38–52.

[VS88] Vanderbei R.J. and Shepp L.A. (1988) A probabilistic model for the time to unravel a strand of DNA. *Comm. Statist. Stochastic Models* 4: 299–314.

[Wei80] Weibel E.R. (1980) *Stereological Methods. Vol 2. Theoretical Foundations.* Academic Press, London.

[Wei82] Weil W. (1982) An application of the central limit theorem for Banach-space-valued random variables to the theory of random sets. *Z. Wahrsch. verw. Gebiete* 60: 203–208.

[Wei83] Weil W. (1983) Stereology: A survey for geometers. In Gruber P.M. and Wills J.M. (eds) *Convexity and Its Applications*, pages 360–412. Birkhäuser, Basel.

[Wei87] Weil W. (1987) Point processes of cylinders, particles and flats. *Acta Appl. Math.* 9: 103–136.

[Wei88] Weil W. (1988) Expectation formulas and isoperimetric properties for non-isotropic Boolean models. *J. Microscopy* 151: 235–245.

[Wei90] Weil W. (1990) Iterations of translative integral formulae and non-isotropic Poisson processes of particles. *Math. Z.* 205: 531–549.

[Wei91] Weil W. (1991) Stochastische Geometrie. Vorlesungskript, Sommersemester.

[Wei93a] Weil W. (1993) The determination of shape and mean shape from sections and projections. *Acta Stereologica* 12(2, Part 1): 73–84.

[Wei93b] Weil W. (1993) Support functions on the convex ring and densities of random sets and point processes. *Suppl. Rend. Circ. Mat. Palermo (2)* 35: 323–344.

[Wei95] Weil W. (1995) The estimation of mean shape and mean particle number in overlapping particle systems in the plane. *Adv. in Appl. Probab.* 27: 102–119.

[Wie86] Wieacker J.A. (1986) Intersections of random hypersurfaces and visibility. *Probab. Theor. Relat. Fields* 71: 405–433.

[Wie89] Wieacker J.A.M. (1989) Geometric inequalities for random surfaces. *Math. Nachr.* 142: 73–106.

[WW84] Weil W. and Wieacker J.A. (1984) Densities for stationary random sets and point processes. *Adv. in Appl. Probab.* 16: 324–346.

[WW87] Weil W. and Wieacker J.A. (1987) A representation theorem for random sets. *Probab. Math. Statist.* 6: 147–151.

[WW93] Weil W. and Wieacker J.A. (1993) Stochastic geometry. In Gruber P.M. and Wills J.M. (eds) *Handbook of Convex Geometry*, pages 1393–1438. Elsevier Sci. Publ., North-Holland, Amsterdam.

[YZ85] Yadin M. and Zacks S. (1985) The visibility of stationary and moving targets in the plane subject to a Poisson field of shadowing elements. *J. Appl. Probab.* 22: 776–786.

[Zäh82] Zähle M. (1982) Random processes of Hausdorff rectifiable closed sets. *Math. Nachr.* 108: 49–72.

[Zäh86] Zähle M. (1986) Curvature measures and random sets, II. *Probab. Theor. Relat. Fields* 71: 37–58.

[Zie89] Ziezold H. (1989) On expected figures in the plane. In Hübler et al.

[HNRW89], pages 105–110.

[Zie94] Ziezold H. (1994) Mean figures and mean shapes applied to biological figures and shape distributions in the plane. *Biometrical J.* 36: 491–510.

Index

WILEY SERIES IN PROBABILITY AND STATISTICS

ESTABLISHED BY WALTER A. SHEWHART AND SAMUEL S. WILKS
Editors
*Vic Barnett, Ralph A. Bradley, Nicholas I. Fisher, J.B. Kadane,
David G. Kendall, David W. Scott, Adrian F. M. Smith, Jozef L. Teugels,
Geoffrey S. Watson*

Probability and Statistics
ANDERSON • An Introduction to Multivariate Statistical Analysis, *Second Edition*
*ANDERSON • The Statistical Analysis of Time Series
ARNOLD, BALAKRISHNAN and NAGARAJA • A First Course in Order Statistics
BACCELLI, COHEN, OLSDER and QUADRAT • Synchronization and Linearity: An Algebra for
    Discrete Event Systems
BARTOSZYNSKI and NIEWIADOMSKA-BUGAJ • Probability and Statistical Inference
BASILEVSKY • Statistical Factor Analysis and Related Methods
BERNARDO and SMITH • Bayesian Statistical Concepts and Theory
BHATTACHARYYA and JOHNSON • Statistical Concepts and Methods
BILLINGSLEY • Convergence of Probability Measures
BILLINGSLEY • Probability and Measure, *Third Edition*
BRANDT, FRANKEN and LISEK • Stationary Stochastic Models
CAINES • Linear Stochastic Systems
CAIROLI and DALANG • Sequential Stochastic Optimization
CHEN • Recursive Estimation and Control for Stochastic Systems
CONSTANTINE • Combinatorial Theory and Statistical Design
COOK and WEISBERG • An Introduction to Regression Graphics
COVER and THOMAS • Elements of Information Theory
CSÖRGÓ and HORVATH • Weighted Approximations in Probability and Statistics
*DOOB • Stochastic Processes
DUDEWICZ and MISHRA • Modern Mathematical Statistics
DUPUIS and ELLIS • A Weak Convergence Approach to the Theory of Large Deviations
ENDERS • Applied Econometric Time Series
ETHIER and KURTZ • Markov Processes: Characterization and Convergence
FELLER • An Introduction to Probability Theory and Its Applications, Volume 1, *Third Edition,*
Revised;   Volume II, *Second Edition*
FREEMAN and SMITH • Aspects of Uncertainty: A Tribute to D.V. Lindley
FULLER • Introduction to Statistical Time Series, *Second Edition*
FULLER • Measurement Error Models
GELFAND and SMITH • Bayesian Computation
GHOSH, MUKHOPADHYAY and SEN • Sequential Estimation
GIFI • Nonlinear Multivariate Analysis
GUTTORP • Statistical Inference for Branching Processes
HALD • A History of Probability and Statistics and Their Applications before 1750
HALL • Introduction to the Theory of Coverage Processes
HAND • Construction and Assessment of Classification Rules
HANNAN and DEISTLER • The Statistical Theory of Linear Systems
*HANSEN, HURWITZ and MADOW • Sample Survey Methods and Theory, 2 Vol Set
HEDAYAT and SINHA • Design and Inference in Finite Population Sampling
HOEL • Introduction to Mathematical Statistics, *Fifth Edition*
HUBER • Robust Statistics
JOHNSON, KOTZ and KEMP • Univariate Discrete Distribution
JUPP and MARDIA • Statistics of Directional Data
JUREK and MASON • Operator-Limit Distributions in Probability Theory
KASS • The Geometrical Foundations of Asymptotic Inference
KAUFMAN and ROUSSEEUW • Finding Groups in Data: An Introduction to Cluster Analysis
KELLY • Probability, Statistics and Optimization

*Now available in a lower priced paperback edition in the Wiley Classics Library

LAMPERTI • Probability: A Survey of the Mathematical Theory, *Second Edition*
LARSON • Introduction to Probability Theory and Statistical Inference, *Third Edition*
LESSLER and KALSBEEK • Nonsampling Error in Surveys
LINDVALL • Lectures on the Coupling Method
McLACHLAN • Discriminant Analysis and Statistical Pattern Recognition
McLACHLAN and KRISHNAN • The EM Algorithm
McNEIL • Epidemiological Research Methods
MANTON, WOODBURY and TOLLEY • Statistical Applications Using Fuzzy Sets
MARDIA • The Art of Statistical Science: A Tribute to G.S. Watson
MARDIA and DRYDEN • Statistical Analysis of Shape
MOLCHANOV • Statistics of the Boolean Model for Practitioners and Mathematicians
MORGENTHALER and TUKEY • Configural Polysampling: A Route to Practical Robustness
MUIRHEAD • Aspects of Multivariate Statistical Theory
OLIVER and SMITH • Inference Diagrams, Belief Nets and Decision Analysis
*PARZEN • Modern Probability Theory and Its Applications
PRESS • Bayesian Statistics: Principles, Models, and Applications
PUKELSHEIM • Optimal Experimental Design
PURI and SEN • Nonparametric Methods in General Linear Models
PURI, VILAPLANA and WERTZ • New Perspectives in Theoretical and Applied Statistics
RAO • Asymptotic Theory of Statistical Inference
RAO • Linear Statistical Inference and Its Applications, *Second Edition*
RAO and SHANBHAG • Choquet-Deny Type Functional Equations and Applications to Stochastic
    Models
RENCHER • Methods of Multivariate Analysis
ROBERTSON, WRIGHT and DYKSTRA • Order Restricted Statistical Inference
ROGERS and WILLIAMS • Diffusions, Markov Processes, and Martingales, Volume I: Foundations,
    Second Edition, Volume II: Itô Calculus
ROHATGI • An Introduction to Probability Theory and Mathematical Statistics
ROSS • Stochastic Processes
RUBINSTEIN • Simulation and the Monte Carlo Method
RUBINSTEIN and SHAPIRO • Discrete Event Systems: Sensitivity Analysis and Stochastic
    Optimization by the Score Function Method
RUZSA and SZEKELY • Algebraic Probability Theory
SCHEFFE • The Analysis of Variance
SEBER • Linear Regression Analysis
SEBER • Multivariate Observations
SEBER and WILD • Nonlinear Regression
SERFLING • Approximation Theorems of Mathematical Statistics
SHORACK and WELLNER • Empirical Processes with Applications to Statistics
SMALL and McLEISH • Hilbert Space Methods in Probability and Statistical Inference
STAPLETON • Linear Statistical Models
STAUDTE and SHEATHER • Robust Estimation and Testing
STOYANOV • Counterexamples in Probability, *Second Edition*
STYAN • The Collected Papers of T.W. Anderson 1943–1985
TANAKA • Time Series Analysis. Nonstationary and Noninvertible Distribution Theory
THOMPSON and SEBER • Adaptive Sampling
WELSH • Aspects of Statistical Inference
WHITTAKER • Graphic Models in Applied Multivariate Statistics
WILLIAMS • Diffusions, Markov Processes, and Martingales, Volume 1. *Second Edition*
YANG • The Construction Theory of Denumerable Markov Processes

Applied Probability and Statistics
ABRAHAM and LEDOLTER • Statistical Methods for Forecasting
AGRESTI • Analysis of Ordinal Categorical Data
AGRESTI • Categorical Data Analysis
AGRESTI • An Introduction to Categorical Data Analysis
ANDERSON and LOYNES • The Teaching of Practical Statistics

*Now available in a lower priced paperback edition in the Wiley Classics Library

ANDERSON, AUQUIER, HAUCH, OAKES, VANDAELE and WEISBERG • Statistical Methods for Comparative Studies

ARMITAGE and DAVID (editors) • Advances in Biometry

*ARTHANARI and DODGE • Mathematical Programming in Statistics

ASMUSSEN • Applied Probability and Queues

*BAILEY • The Elements of Stochastic Processes with Applications to the Natural Sciences

BARNETT and LEWIS • Outliers in Statistical Data, *Third Edition*

BARTHOLOMEW, FORBES, and McLEAN • Statistical Techniques for Manpower Planning, *Second Edition*

BATES and WATTS • Nonlinear Regression Analysis and Its Applications

BECHOFER, SANTNER and GOLDSMAN • Design and Analysis of Experiments for Statistical Selection, Screening and Multiple Comparisons

BELSLEY • Conditioning Diagnostics: Collinearity and Weak Data in Regression

BELSLEY, KUH and WELSCH • Regression Diagnostics: Identifying Influential Data and Sources of Collinearity •

BERNARDO and SMITH • Bayesian Theory

BERRY, CHALONER and GEWEKE • Bayesian Analysis in Statistics and Econometrics Essays in Honor of Arnold Zellner

BHAT • Elements of Applied Stochastic Processes, *Second Edition*

BHATTACHARYA and WAYMIRE • Stochastic Processes with Applications

BIEMER, GROVES, LYBERG, MATHIOWETZ and SUDMAN • Measurement Errors in Surveys

BIRKES and DODGE • Alternative Methods of Regression

BLOOMFIELD • Fourier Analysis of Time Series: An Introduction

BOLLEN • Structural Equations with Latent Variables

BOULEAU • Numerical Methods for Stochastic Processes

BOX • R.A.Fisher, the Life of a Scientist

BOX and DRAPER • Empirical Model-Building and Response Surfaces

BOX and DRAPER • Evolutionary Operation: A Statistical Method for Process Improvement

BOX, HUNTER and HUNTER • Statistics for Experimenters: An Introduction to Design, Data Analysis, and Model Building

BROWN and HOLLANDER • Statistics: A Biomedical Introduction

BUCKLEW • Large Deviation Techniques in Decision, Simulation, and Estimation

BUNKE and BUNKE • Non-linear Regression, Functional Relations and Robust Methods: Statistical Methods of Model Building

CHATTERJEE and HADI • Sensitivity Analysis in Linear Regression

CHATTERJEE and PRICE • Regression Analysis by Example, *Second Edition*

CLARKE and DISNEY • Probability and Random Processes: A First Course with Applications, *Second Edition*

COCHRAN • Sampling Techniques, *Third Edition*

*COCHRAN and COX • Experimental Designs, *Second Edition*

CONOVER • Practical Nonparametric Statistics, *Second Edition*

CORNELL • Experiments with Mixtures, Designs, Models, and the Analysis of Mixture Data, *Second Edition*

COX • A Handbook of Introductory Statistical Methods

*COX • Planning of Experiments

COX, BINDER, CHINNAPPA, CHRISTIANSON, COLLEDGE, and KOTT • Business Survey Methods

CRESSIE • Statistics for Spatial Data, *Revised Edition*

DANIEL • Applications of Statistics to Industrial Experimentation

DANIEL • Biostatistics: A Foundation for Analysis in the Health Sciences, *Sixth Edition*

DANIEL   Fitting Equations into Data: Computer Analysis of Multifactor Data, *Second Edition*

DAVID • Order Statistics, *Second Edition*

*DEGROOT, FIENBERG and KADANE • Statistics and the Law

*DEMING • Sample Design in Business Research

DILLON and GOLDSTEIN • Multivariate Analysis: Methods and Applications

DOWDY and WEARDEN • Statistics for Research, *Second Edition*

DRAPER and SMITH • Applied Regression Analysis, *Second Edition*

DUNN • Basic Statistics: A Primer for the Biomedical Sciences, *Second Edition*

KALBFLEISCH and PRENTICE • The Statistical Analysis of Failure Time Data
KASPRZYK, DUNCAN, KALTON and SINGH • Panel Surveys
KHURI • Advanced Calculus with Applications in Statistics
KISH • Statistical Design for Research
*KISH • Survey Sampling
KOTZ • Personalities
KOVALENKO, KUZNETZOV and PEGG • Mathematical Theory of Reliability of Time-dependent
    Systems with Practical Applications
LAD • Operational Subjective Statistical Methods: A Mathematical, Philosophical and Historical
    Introduction
LANGE, RYAN, BILLARD, BRILLINGER, CONQUEST, and GREENHOUSE • Case Studies in
    Biometry
LAWLESS • Statistical Models and Methods for Lifetime Data
LEE • Statistical Methods for Survival Data Analysis, *Second Edition*
LePAGE and BILLARD • Exploring the Limits of Bootstrap
LESSLER and KALSBEEK • Nonsampling Error in Surveys
LEVY and • LEMESHOW • Sampling of Populations: Methods and Applications
LINHART and ZUCCHINI • Model Selection
LITTLE and RUBIN Statistical Analysis with Missing Data
LYBERG • Survey Measurement
McLACHLAN • Discriminant Analysis and Statistical Pattern Recognition
McLACHLAN and KRISHNAN • The EM Algorithm and Extensions
McNEIL • Epidemiological Research Methods
MAGNUS and NEUDECKER • Matrix Differential Calculus with Applications in Statistics and
    Econometrics
MALLER and ZHOU • Survival Analysis with Long Term Survivors
MALLOWS • Design, Data, and Analysis by Some Friends of Cuthbert Daniel
MANN, SCHAFER, and SINPURWALLA • Methods for Statistical Analysis of Reliability and Life
    Data
MASON, GUNST, and HESS • Statistical Design and Analysis of Experiments with Applications to
    Engineering and Science
MILLER • Survival Analysis
MONTGOMERY and MYERS • Response Surface Methodology: Process and Product in Optimization
    Using Designed Experiments
MONTGOMERY and PECK • Introduction to Linear Regression Analysis, *Second Edition*
MORGENTHALER and TUKEY • Configural Polysampling
MYERS and MONTGOMERY • Response Surface Methodology
NELSON • Accelerated Testing, Statistical Models, Test Plans, and Data Analyses
NELSON • Applied Life Data Analysis
OCHI • Applied Probability and Stochastic Processed in Engineering and Physical Sciences
OKABE, BOOTS, and SUGIHARA • Spatial Tesselations: Concepts and Applications of Voronoi
    Diagrams
PANKRATZ • Forecasting with Dynamic Regression Models
PANKRATZ • Forecasting with Univariate Box-Jenkins Models: Concepts and Cases
PORT • Theoretical Probability for Applications
PUKELSHEIM • Optimal Design of Experiments
PUTERMAN • Markov Decision Processes: Discrete Stochastic Dynamic Programming
RACHEV • Probability Metrics and the Stability of Stochastic Models
RADHAKRISHNA RAO and SHANBHAG • Choquet-Deny Type Functional Equations with
    Applications to Stochastic Models
RÉNYI • A Diary on Information Theory
RIPLEY • Spatial Statistics
RIPLEY • Stochastic Simulation
ROSS • Introduction to Probability and Statistics for Engineers and Scientists
ROUSSEEUW and LEROY • Robust Regression and Outlier Detection
RUBIN • Multiple Imputation for Nonresponse in Surveys
RUBINSTEIN and SHAPIRO Discrete Event Systems: Sensitivity Analysis and Stochastic
    Optimization by the Score
RYAN • Modern Regression Methods

*Now available in a lower priced paperback edition in the Wiley Classics Library

RYAN • Statistical Methods for Quality Improvement
SCHOTT • Matrix
SCOTT • Multivariate Density Estimation: Theory, Practice, and Visualization
SEARLE • Linear Models
SEARLE • Linear Models for Unbalanced Data
SEARLE • Matrix Algebra Useful for Statistics
SEARLE, CASELLA and McCULLOCH • Variance Components
SKINNER, HOLT, and SMITH • Analysis of Complex Surveys
STOYAN, KENDALL, and MECKE • Stochastic Geometry and Its Applications, *Second Edition*
STOYAN and STOYAN • Fractals, Random Shapes and Point Fields: Methods of Geometrical
  Statistics
THOMPSON • Empirical Model Building
THOMPSON • Sampling
TIERNEY • LISP-STAT: An Object-Oriented Environment for Statistical Computing and Dynamic
  Graphics
TIJMS • Stochastic Models: An Algorithmic Approach
TITTERINGTON, SMITH and MARKOV • Statistical Analysis of Finite Mixture Distributions
UPTON and FINGLETON • Spatial Data Analysis by Example, Volume 1: Point Pattern and
  Quantitative Data
UPTON and FINGLETON • Spatial Data Analysis by Example, Volume II: Categorical and
  Directional Data
VAN RIJKEVORSEL and DE LEEUW • Component and Correspondence Analysis
WEISBERG • Applied Linear Regression, *Second Edition*
WESTFALL and YOUNG • Resampling-Based Multiple Testing: Examples and Methods for p-Value
  Adjustment
WHITTLE • Optimization Over Time: Dynamic Programming and Stochastic Control, Volume 1 and
  Volume II
WHITTLE • Systems in Stochastic Equilibrium
WONNACOTT and WONNACOTT • Econometrics, *Second Edition*
WONNACOTT and WONNACOTT • Introductory Statistics, *Fifth Edition*
WONNACOTT and WONNACOTT • Introductory Statistics for Business and Economics, *Fourth
  Edition*
WOODING • Planning Pharmaceutical Clinical Trials: Basic Statistical Principles
WOOLSON • Statistical Methods for the Analysis of Biomedical Data
*ZELLNER • An Introduction to Bayesian Inference in Econometrics

Tracts on Probability and Statistics
BILLINGSLEY • Convergence of Probability Measures
KELLY • Reversibility and Stochastic Networks